Plant-Pest Interactions

Plant-Pest Interactions

Edited by:

Kiran Abasaheb More

Delve
PUBLISHING

www.delvepublishing.com

Plant-Pest Interactions

Kiran Abasaheb More

Delve Publishing
224 Shoreacres Road
Burlington, ON L7L 2H2
Canada
www.delvepublishing.com
Email: orders@arclereducation.com

ISBN: 978-1-77469-190-8 (Hardcover)

ABOUT THE EDITOR

Dr. Kiran Abasaheb More, born in Bhogaon (Devi) Tahsil Jintur of Parbhani district of Maharashtra, India, in 1978, completed M.Sc in Botany, at Shree Shivaji College Parbhani, Maharashtra, Affiliated with Swami Ramanand Teerth Marathwada University Nanded, Maharashtra. The author also completed his Ph.D in 2010 in field of Taxonomy of Angiosperms and he joined the department of Botany, Yashwantrao Chavan Arts and Science Mahavidyalaya, Mangrulpir as an Assistant Professor. He is the Chief Editor of Research Journal Multilogic in science (Journal for Applied science), He edited more than 35 issues of research journals and authored 25 research papers. Dr. More is awarded by Young scientist award of Science and tech society for Integrated ruler improvement Telangana , and also awarded by Excellence in Teaching Genesis Arban &Ruler Devolopment Society at the occasion of ICAAASTSD International conference 2018. He is the editor of proceeding of International conference on Advances in Agriculture and Allied science technologies for sustainable Development Society Special issue Jan 2018 and he is a Life Member of "GUARD Society". He also chaired a number of National and International Conferences, he is also a Member of Botany teacher association of Sant Gadge baba Amravati University Amravati, Maharashtra, India.

TABLE OF CONTENTS

LIST OF FIGURES

LIST OF TABLES

LIST OF ABBREVIATIONS

ALP	Alkaline Phosphatase
AMF	Arbuscular Mycorrhizal Fungus
AOS	Allene Synthase
CAM	Calmodulin
CDPK	Calcium-Dependent Protein Kinases
CDPK	Calcium-Based Protein Kinases
DB	Dioscorea batatas
EST	Expressed Collection Tag
FISH	Fluorescence in Situ Hybridization
GOX	Glucose Oxidase
HDL	High-Density Lipoproteins
HEL	Hevein Like
HIPV	Herbivore Induced Plant Volatiles
HPL	Hydro PeroxideLyase
IGL	Indole-Three-glycerol Phosphate Lyase
ISR	Induced Systemic Resistance.
JA	Jasmonic Acid
LOX	Lipoxygenase
MAPK	Mitogen-Activated Protein Kinase
MARK	Mitogen-Activated Protein Kinase
MEBA	Methyl Benzoate
MEJA	Methyl Jasmonate
MESA	Methyl Salicylate
NGS	Next Generation Sequencing
OUT	Operational Taxonomic Units
PAL	Phenylalanine Ammonia-Lyase
PCR	Polymerase Chain Reaction
PGPR	Plant Growth Promoting Rhizobacteria

PI	Proteinase inhibitor
ROS	Reactive Oxygen Species
SA	Salicylic Acid
TFs	Transcription Factor Regulation

GLOSSARY

A

Abiotic–In biology and ecology, abiotic components or abiotic factors are non-living chemical and physical parts of the environment that affect living organisms and the functioning of ecosystems. Abiotic factors and the phenomena associated with them underpin biology as a whole.

Adaptation– In biology, adaptation has three related meanings. Firstly, it is the dynamic evolutionary process that fits organisms to their environment, enhancing their evolutionary fitness. Secondly, it is a state reached by the population during that process.

Alkaloids – Alkaloids are a class of basic, naturally occurring organic compounds that contain at least one nitrogen atom. This group also includes some related compounds with neutral and even weakly acidic properties. Some synthetic compounds of similar structures may also be termed alkaloids.

Amino Acid – Amino acids are the building blocks of proteins and play an important role in body functions. They are needed for vital processes like cell building and the synthesis of hormones and neurotransmitters (brain chemicals).

Arabidopsis -Arabidopsis is a genus in the family Brassicaceae. They are small flowering plants related to cabbage and mustard. This genus is of great interest since it contains thale cress, one of the model organisms used for studying plant biology and the first plant to have its entire genome sequenced.

Archaebacteria– Archaebacteria are a group of microorganisms considered to be an ancient form of life that evolved separately from the bacteria and blue-green algae, and they are sometimes classified as a kingdom.

B

Beetle- Beetles are a group of insects that form the order Coleoptera, in the superorder Endopterygota. Their front pair of wings are hardened into wing-cases, elytra, distinguishing them from most other insects.

Benzyl– In organic chemistry, benzyl is the substituent or molecular fragment possessing the structure $C_6H_5CH_2$–. Benzyl features a benzene ring attached to a CH_2 group.

Biosynthesis – Biosynthesis is a multi-step, enzyme-catalyzed process where substrates are converted into more complex products in living organisms. In biosynthesis, simple compounds are modified, converted into other compounds, or joined together to form macromolecules. This process often consists of metabolic pathways.

C

Carbon Cycle – The carbon cycle is the biogeochemical cycle by which carbon is exchanged among the biosphere, pedosphere, geosphere, hydrosphere, and atmosphere of the Earth. Carbon is the main component of biological compounds as well as a major component of many minerals such as limestone.

Chemical–A chemical substance is a form of matter having a constant chemical composition and characteristic properties. Some references add that a chemical substance cannot be separated into its constituent elements by physical separation methods, i.e., without breaking chemical bonds.
Chickpea–The chickpea or chick pea is an annual legume of the family Fabaceae, subfamily Faboideae. Its different types are variously known as gram or Bengal gram, garbanzo or garbanzo bean, or Egyptian pea. Chickpea seeds are high in protein.
Compound – A chemical compound is a chemical substance composed of many identical molecules composed of atoms from more than one element held together by chemical bonds. A molecule consisting of atoms of only one element is therefore not a compound.
Crop – A crop is a plant or animal product that can be grown and harvested extensively for profit or subsistence. Crops may refer either to the harvested parts or to the harvest in a more refined state. Most crops are cultivated in agriculture or aquaculture. A crop may include macroscopic fungus, or alga.

D

Digestibility – Digestibility refers to the amount of nutrient absorbed by the individual and is generally calculated as the amount of nutrient consumed minus the amount of nutrient retained in the feces.
Drought– A drought is an event of prolonged shortages in the water supply, whether atmospheric, surface water or groundwater. A drought can last for months or years, or maybe declared after as few as 15 days.

E

Ecology–Ecology is the study of the relationships between living organisms, including humans, and their physical environment. Ecology considers organisms at the individual, population, community, ecosystems, and biosphere level.
Ectomycorrhiza – An ectomycorrhiza is a form of symbiotic relationship that occurs between a fungal symbiont, or mycobiont, and the roots of various plant species. The mycobiont is often from the phyla Basidiomycota and Ascomycota, and more rarely from the Zygomycota.
Elicitors– Elicitors in plant biology are extrinsic or foreign molecules often associated with plant pests, diseases or synergistic organisms. Elicitor molecules can attach to special receptor proteins located on plant cell membranes.
Epicuticular wax- Epicuticular wax is a coating of wax covering the outer surface of the plant cuticle in land plants. It may form a whitish film or bloom on leaves, fruits and other plant organs. The main functions of the epicuticular wax are to decrease surface wetting and moisture loss.
Ethylene – Ethylene is a hydrocarbon that has the formula C_2H_4 or $H_2C{=}CH_2$. It is a colorless flammable gas with a faint "sweet and musky" odor when pure. It is the simplest alkene. Ethylene is widely used in the chemical industry, and its worldwide production exceeds that of any other organic compound.

Eukaryotes -Eukaryotes are organisms whose cells have a nucleus enclosed within a nuclear envelope. Eukaryotes belong to the domain Eukaryote or Eukarya; their name comes from Greek.

F

Fatty Acid -In chemistry, particularly in biochemistry, a fatty acid is a carboxylic acid with a long aliphatic chain, which is either saturated or unsaturated. Most naturally occurring fatty acids have an unbranched chain of an even number of carbon atoms, from 4 to 28.

Flavonoids -Flavonoids are a class of polyphenolic secondary metabolites found in plants, and thus commonly consumed in diets. Chemically, flavonoids have the general structure of a 15-carbon skeleton, which consists of two phenyl rings and a heterocyclic ring. This carbon structure can be abbreviated C6-C3-C6.

Fungal– A fungus that invades the tissue can cause a disease that's confined to the skin, spreads into tissue, bones and organs or affects the whole body.

G

Galactomannan – Galactomannan (GM) is a polysaccharide cell wall component of Aspergillus sp. that is released into the blood during fungal growth in tissue.

Genomes– In the fields of molecular biology and genetics, a genome is all the genetic material of an organism. It consists of DNA. The genome includes both the genes and the noncoding DNA, as well as mitochondrial DNA and chloroplast DNA. The study of the genome is called genomics.

Genomics- Genomics is an interdisciplinary field of biology focusing on the structure, function, evolution, mapping, and editing of genomes. A genome is an organism's complete set of DNAS, including all of its genes.

Glycerol -Glycerol is a simple polyol compound. It is a colorless, odorless, viscous liquid that is sweet-tasting and non-toxic. The glycerol backbone is found in those lipids known as glycerides. Due to having antimicrobial and antiviral properties, it is widely used in FDA-approved wound and burn treatments.

H

Hazard -A hazard is a source or a situation with the potential for harm in terms of human injury or ill-health, damage to property, damage to the environment, or a combination of these. An unwanted event is a situation or condition where there is a loss of control of the hazard that leads to harm.

Herbivore – A herbivore is an animal anatomically and physiologically adapted to eating plant material, for example, foliage or marine algae, for the main component of its diet. As a result of their plant diet, herbivorous animals typically have mouthparts adapted to rasping or grinding.

of plant origin, pest attack on plant and activation of MAPK cascade, early signal transduction and role of kinases, the role of MAPKs in plant hormones signaling.

The seventh chapter explains the role of Insecticidal Proteins and Their Importance in controlling Plant Pests. This chapter also emphasizes destruction caused by insect pests, the role of insecticidal protein, advances in BT toxins insecticidal capacity, plant lectins, mode of action of plant lectins, and safety assessment of lectin.

The last chapter of this book sheds light on the role of microbes in plant-insect interactions. This chapter also mentions the importance of microbial diversity involved in plant-pest interaction, beneficial aspects of microbes-plant-pest interactions and visual cues, harmful aspects of microbes-plant-pest interactions and visual cues, bacterial and fungal interactions, microbes and horizontal gene transfer in microbe-plant-insect interaction, and the impact of microbes on insect reproduction.

This book has been designed to suit the knowledge and pursuit of the researcher and scholars and to empower them with various aspects of plant-pest interactions, so that they are updated with the information. I hope that the readers find the book explanatory and insightful and that this book is referred by scholars across various fields.

Chapter 1

Introduction to Plant Pest Interactions

CONTENTS

Extensive studies on Plant-Pest interactions on cellular as well as on organismic levels have opened new doors through which we can understand this relationship more thoroughly and can produce better quality plants and food crops.

Plant and Pest interactions are based upon the fact that pests cause many harmful diseases among plants through some common disease-causing characteristics. Plant-Pest interactions are also influenced by physical as well as chemical factors. The pest interacts with plants in different ways and the way of their interaction is quite attractive as it acts as a road map to get improved quality of plants and crops.

1.1. INTRODUCTION

The evolution of plants altered the terrestrial environment into an extremely valuable resource for the herbivore community. In natural ecosystems, insects and plants are just some of the living organisms that are constantly interacting in an intricate way.

These two organisms are strictly allied since insects have numerous advantageous activities comprising pollination and defense while plants offer shelter, oviposition sites and food, the three central factors necessitated for insect proliferation (Panda and Khush, 1995).

On the other side, contingent on the intensity of insect attack, herbivores might be greatly harmful to plants leading to death. Plant-insect interaction is a dynamic system, subjected to repetitive variation and alteration.

In order to lessen the attack by insects, plants developed distinctive defense mechanisms comprising physical and chemical barriers such as the induction of defensive proteins (Haruta et al., 2001), volatiles that entice predators of the insect herbivores (Birkett et al., 2000), references herein; (Kliebenstein et al., 2001), secondary metabolites (Baldwin, 2001) and trichome density (Fordyce and Agrawal, 2001).

In parallel, insects advanced strategies to get over through plant barriers such as detoxification of toxic compounds (Scott and Wen, 2001), sequestration of poison (Nishida, 2002 and references herein, avoidance mechanisms (Zangerl, 1990) and change of gene expression pattern (Silva et al., 2001).

Plants yield chemicals for defense objectives in two distinctive ways; first, as constitutive substances to deter herbivores through direct toxicity or by curtailing the digestibility of plant tissues and second, as inducible substances synthesized in reply to tissue damage by herbivores. These tactics can avert the majority of the herbivores although there are few insects that are able to acclimate to plant species.

Secondary metabolites perform valuable functions for the plant acting either in a constitutive or inducible manner. Some compounds are plant growth regulators while others act as chemical signals in antibiosis agents, the ecosystem, storage and conveyance of carbon and nitrogen molecules that are directly involved in the plant primary metabolism (Panda and Khush, 1995). Secondary plant compounds are entailed in plant defense against insect herbivores acting as feeding inhibitors, insect repellents, and/ or toxins.

In this chapter, we have categorized these toxic compounds into chemical-derived substances and protein-derived molecules. Glucosinolates are an instance of secondary metabolites involved in insect/plant interactions. This class of molecules varies qualitatively in Arabidopsis plants, making diverse combinations in relation to changing herbivory or other selective pressures (Kliebenstein et al., 2001). The emission of volatile compounds is another imperative mechanism impacting the behavior of insects searching for food.

It is generally seen that odors from plants are one of the primary cues that insects use to find the host plant. For instance, cabbage seed weevil looks to be orientated by an intricate mixture of host plant volatiles (Bartlet et al., 1993).

The existence of chemical volatile compounds in plants specifies a double meaning. First, they can deter a broad range of potential herbivores because of the nature of toxic compounds released in the air. Secondly, they have the property of enticing a small number of focused pest species and of acting as an indirect plant defense mechanism by enticing other insects that prey on or parasitize the herbivores (Birkett et al., 2000).

According to Kessler and Baldwin (2001), the volatile cocktail unconstrained by tobacco plants entices predatory bugs to tobacco hornworm eggs and feeding larvae severely increasing the predation rates. Moreover, these released volatiles lessen oviposition rates from adult moths as adults avoid plants on which predators are probable to be present, lessening herbivore loads by 90 percent.

Plant volatiles may also perform as signals between plants, where volatiles from a damaged tissue stimulate a defense response in adjoining undamaged plants (Paré and Tumlinson, 1999). According to Paré and Tumlinson (1999), each plant yields a herbivore-specific blend of volatile components in relation to a particular elicitor from a specific herbivore species feeding on the leaves.

This greatly suggests the coevolution between herbivores, plants, and their natural enemies. Wounds created by herbivory can be resourcefully used as points of invasion by microorganisms and signify an augmented risk of future pathogen attack (Paul et al., 2000).

But the induced resistance lets full expression of the plant's potential to bear either disease or herbivory. This has been experimentally shown by the observation that insect feeding persuades the production of phytoalexins that have antimicrobial properties (Ananthakrishnan, 1999).

In addition to this, intact cells surrounding areas of damaged tissue form physical barriers to confine pathogen invasion by strengthening the cell wall, sealing the wound site, or segregating the cells from their neighbors (de Bruxelles and Roberts, 2001).

According to Haruta et al. (2001), *Populus tremuloides* engages in a broader range of defensive strategies against insect herbivores comprising both protein and phytochemical-based components. The multifaceted array of defense mechanisms has been embraced in most plant-herbivore interactions. Although the production of various plant compounds can be very effective to curtail insect damage, it is costly in most cases.

During an herbivore attack, a plant might suffer two sorts of fitness consequences: tissue loss and large investments needed for resistance. These costs are an imperative component for the evolution of resistance because an evolutionary equilibrium should be recognized at intermediate levels of herbivory fitness and reduction (Paul et al., 2000).

Thus, plants might be able to avoid allocation costs by evolving more effective alternative biochemical pathways. Inducible defenses can be a substitute to curtail fitness costs since they decrease the plant's expenditures by letting it invest in defense when necessary, and to avoid costly distribution to defense when herbivores are not existent (Agrawal, 2000). Another alternative is used by some plants and is the suppression of the early and expensive defense events at the time of attack recognition (Baldwin, 2001).

Insect-plant interaction mainly denotes the activities of two sorts of organisms: insects that seek out and utilize plants for food, egg-laying sites, and/or, shelter, and the plants that offer those resources. These interactions are generally assessed from the plant's perspective, and a principal broad research question is: "How do the activities of the insect affect plant growth and development?" The interactions can be advantageous to both the plant and the insect, as illustrated by pollination. During pollination, an insect moving within a flower to attain nectar may transfer pollen either within that flower or among other flowers on that plant.

Other relationships between plants and insects can be damaging to the plant but advantageous to the insect (e.g., herbivory, or feeding upon the plant). Plant-feeding insect species are abundant, comprising more than one-quarter of all macroscopic organisms. Although most plant parts are fed upon by insect herbivores, the insect herbivores predominantly exhibit specificity with reference to the plant species and the plant part on which they will feed.

Some instances of imperative insect herbivores across the world on cultivated crops include: aphids on cereal crops, diamondback moth larvae (immatures) on members of the cabbage family, and larvae of the moth genus *Heliothis* on a wide-ranging of plants, involving cotton. Apart from the direct effects of herbivory, insects can be destructive to plants by acting as vectors (carriers) of pathological microorganisms, conveying the organisms when the insects feed on the plants.

1.1.1. Interactions in Agricultural Settings

In a way to avert substantial losses of agricultural crops to herbivores, both in the field and following harvest, some form of insect population control is generally needed; some crops may need protection from more than one insect herbivore.

Under traditional farming methods in the industrialized world, insecticides are applied to agricultural fields to control insect pests. Usually, more than one sort of insecticide and/or more than one treatment will be applicable in a single crop cycle. The type of control method used for a specific crop/insect combination in part be contingent upon the understanding of the insect and its use of a specific crop plant. Research into novel aspects of insect-plant interaction may offer improved alternatives for controlling insect pest populations.

For example, recent research inspecting the effects of moth larvae feeding on corn has confirmed that after herbivore damage, corn plants release a new intricate of **odorants** into the air, and that some of these molecules are appealing to parasitic wasps. The parasitic wasps then seek out and parasitize the larvae feeding on the corn plants. These odorants can be used to help control moth damage on corn.

A very distinctive view of insect-plant interaction emphasizes the use of insects as biological control agents for weeds and takes benefit of the fact that insects can feed detrimentally on plants. A well-known instance of insect control of weeds occurred in Australia when prickly pear cacti were monitored by the cactus moth from Argentina, *Cactoblástis cactòrum* (Berg), an insect herbivore imported for that determination.

Some insect-plant relationships can be tracked through the fossil record, as some fossilized leaves exhibit an indication of ancient herbivory that arisen preceding the fossilization of the plant material. Other insect-plant relationships endure to grow as insect species integrate novel host plants into their diets and plants progress a new defensive compound. The dynamic variety and nature of these interactions offer much opportunity to enhance our understanding of the **physiology** of both sorts of organisms, interactions between them, and evolutionary and ecological processes.

Pests and pathogens are always seen to cause damage to plants and crops since time immemorial. Pest interferes with the plant growth and development, attacks on their cellular machinery and affects their genetic makeup. According to surveys, damage to the plants caused by pests accounts for approximately 35-40% of global annual crop destruction (Follett, 2017). The loss of crops by pesticide is much greater in the countries that are poorly developed and lack the modern techniques to manage crops properly. Humans always seem to discover the ways through which they can control pests leading to the production of pesticides. However, these pesticides lead to the fear of pesticide resistance in a crop, as well as the negative impact of these pesticides on human health.

Most insects in nature usually get their food from plants and their way of ingesting also varies within different varieties of insects. However, plants also have evolved to have advanced resistive mechanisms against such insects and interact with them in many fascinating ways without being killed by those pests (Painter, 1951). Pest and their attack on important sensitive food crops are one of the major concerns of farmers as it can lead to destruction of an entire area of crop. The present-day plant crops emerged as the result of continuous interactions with pests that evolved them to become more resistive to harmful hazardous disease-causing pests. It leads to the evolution of more tolerant plant species against such pests through the process of natural selection having more strong defensive mechanisms against harmful pests (Pedigo, 1999). The discovery of modern biotechnological techniques such as GMOs with better and improved characteristics has increased the crop susceptibility to damage by pests and environmental factors more frequently (Chen, Gols and Benrey, 2015). One of the successful approaches of recent advancement in science is the study of plant-pest interactions with response to fast-paced molecular and chemical biology. It is the rapidly growing approach that is being applied in recent agricultural practices to gain a large number of economic benefits by producing more valuable plant crops.

1.2. HISTORY OF PLANT PEST INTERACTIONS

The Plant-pest interaction is not a new approach but has been known since ancient times. It is common knowledge that insects have originated in the Devonian period, and evolved with plants through co-evolution in order to make a strong relationship with plants termed as plant-pest interaction.

Charles Darwin was the first one who described the plant-pest relationship through the process of coevolution. He observed that orchid flowers co-evolved with their insect pollinators to allow for successful cross-pollination. Dethier was very keen to work on the species of *Papilio* and *Umbelliferae*. He emphasized the importance of the chemicals that induce attraction between plants and pest to establish the successful plant-pest relationship.

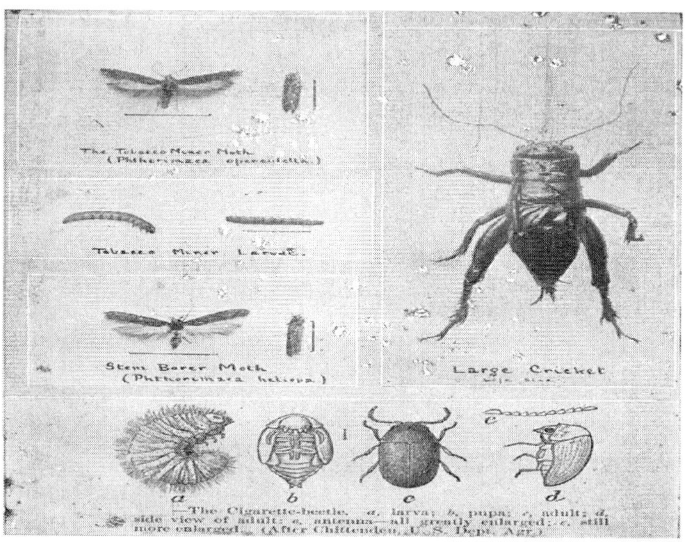

Figure 1.1. History of Tobacco Pests Southern Rhodesia.

Source: Image by environmentandsociety.org

Fritz and his colleagues in 1992 wrote comprehensive books on plant interactions with organisms feeding on them. The major focus of their study was the resistance of plants against harmful pests with reference to ecological and evolutionary points of view. Hopkins' Host Selection Principle was given by Hopkin when he was investigating the bark of plan scolytif and timber beetles. According to this principle when a species can breed on more than two hosts then it prefers to be bred on the host species on which it became adapted (Hopkins, 2015).

R.H. Painter in 1951, for the first time, gave the concept of the resistance mechanism of plants against the insects. He further investigated the resistance mechanism through which a plant protects itself against the disease-causing agents, particularly insects. These mechanisms through which plant provides resistance against the pests are the production of antibiotics, induction of tolerance and non-preference mechanisms.

1.3. PLANT HEALTH AND DISEASE

A plant is healthy if its genetic makeup remains the same and there are no mutations of deleterious nature present within its genome. The ideal plant health with no mutation in its genetic material is further supported by the presence of ideal environmental factors such as soil, suitable weather, light, and the presence of useful microbes within the soil. The healthy plant can perform its normal functions without any difficulty. The tasks that a plant must perform are absorptions of water and essential nutrients from the soil through their roots, food production through the process of photosynthesis, the transport of photosynthetic products to the entire plant body, making offspring through the process of reproduction and control of water movement through the process of transpiration within the plant's roots and leaves. When a plant is seen to be performing all its functions normally it is healthy. A healthy plant tends to grow, multiply normally and produce the ideal amount of crop yield.

But when a plant is not able to perform its normal functions due to biotic or abiotic factors it becomes diseased and lead to many complications as now the plant is not able to perform its normal functions required for its optimal growth. The infected plant is then examined to check if it has certain disease symptoms caused by the insects or pests. It further gives information about what kind of measures should be taken in order to cure the plant and to stop the disease from spreading to the other parts of the same plant as well as to the other plants (Guest, 2016).

1.4. PLANT PEST INTERACTION AND PRODUCTION OF VOLATILE COMPOUNDS

Our nature works in combination with both abiotic and biotic factors in order to sustain life and energy is continuously flowing between living and nonliving components of the ecosystems. Plants are living organisms that use their own food through the process of photosynthesis by using their photosynthetic machinery in the chloroplast. Plants not only produced food for themselves but insects and other organisms are also dependent on plants to fulfill their nutritional requirements. The agricultural crops are the most important parts of the plant ecosystems that continuously interact with pests through many complex interactions. One of the modes of interaction of the plant and pest relationship is the production of volatile compounds that may be produced in different plant parts like leaves, fruits and flowers and stored in the resin duct or in glandular trichomes of the plants (Gershenzon et al., 2000).

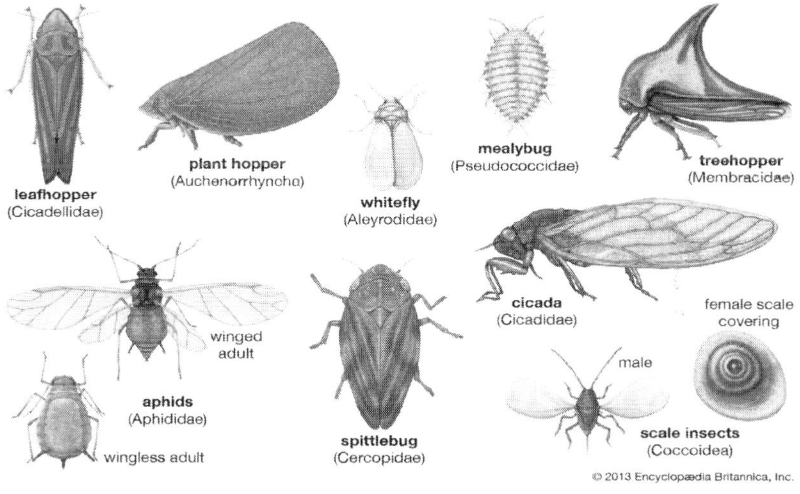

Figure 1.2. Different Types of Plant Pests.

Source: Image by pxfuel.com

The plants under normal conditions usually produce volatiles in fixed ratios that can be changed when plants face some stressful conditions in the form of environmental stresses or some pest attack. There are two types of plant-based volatiles compounds that are produced within plants and release during damage by pests or insects.

- Constitutive plant-based volatile compounds
- Inducible plant-based volatile compounds

These are the following important constitutive plant-based volatile compounds are green leaf volatiles:

- C6 aldehydes volatiles,
- C6 alcohols volatiles
- Acetate volatiles
- (z)-3-hexenol, hexenal (produced in *Arabidopsis thaliana* during herbivore stress condition)

Plant produces these volatiles on a regular basis but the amounts of their production are rapidly increased when plant face stress as the result of herbivore attack.

The second group of herbivores induced plant volatiles (HIPV's) includes (z)-3- hexenyl acetate, (z)-3-hexenyl butyrate, indole and various terpenoids such as α-farnesene, β-farnesene, β-ocimene and linalool (Rose and Tumlison, 2004).

Few important examples of induced plant volatiles that are produced by herbivore attack are as follows:

- Indole
- (z)-3- hexenyl acetate
- (z)-3-hexenyl butyrate
- Terpenoids such as α-farnesene and linalool (Rose and Tumlison, 2004).

The pest interacted with plant volatile production by secreting certain elicitors that activate the plant's defense mechanism as can be seen in the case of caterpillars feeding on the leaves of the plants. Larvae of *Spodoptera littoralis* were seen to contain oral secretions to initiate the elicitation-based-defense response from the host plants. Larvae of *Spodoptera litura* attack and feed on the leaves of soybeans to activate the plant defense mechanism by eliciting the production of flavones along with iso flavone to trigger the ability of the plant to survive (Nebapure and Sagar, 2015).

Several studies also proposed that when a pest lays eggs on the plant, their deposition leads to the secretion of volatile compounds by the plant in order to initiate their defense mechanism. The secretion of volatile compounds not only kills the pest eggs but also kills the enemies of parasitic origin that attracted to pest eggs to generate its defense response. For example, when the elm leaf battle deposited their eggs on the leave of elm, elm plant started releasing volatile compound that attracts beetle parasite *Oomyzus galerucivorous* in order to get relief from the eggs of *Xanthogaleruca luteola*.

1.4.1. Role of Plant Volatiles against Enemies of Pest

When a plant gets attacked by a pest, the pest contains elicitors that initiate volatile compound synthesis. The production of volatile compounds not only inhibits egg deposition of pests but also attracts the naturally occurring pest animals in order to get rid of them is known as the systematic release of plant volatiles induced by plant herbivory. For example, cabbage plant roots are eaten by larvae of root fly *Delia radicum* (L.) at lead to the production of volatile compounds that attracts *Trybliographa rapae. Trybliographa rapa* is an endoparasitoid of root fly in order to free the plant roots from the parasite (Guest, 2016). Through these mechanisms of volatile compound production by plants get rid of pests by attracting insect pest naturally occurring enemies.

Pieris rapae (L.) attacks*Arabidopsis thaliana*, blend of various aliphatics terpenoids compounds along with methyl salicylate are released by the Arabidopsis plant in order to attract larval parasitoid *Cotesia rubecula* to avoid the deposition of pest eggs (Poecke et al., 2001).

1.5. DEVELOPMENT OF PLANT PROTECTION STRATEGIES AGAINST PESTS

Complex and delicate mechanisms are usually used by insects to establish a relationship with their host plants. Additionally, the physiological diversifications during an exceedingly diverse condition of the plant in response to herbivory, which attracts parasitoids and predators, it becomes imperative to keep a check on such interactions in the agricultural ecosystem.

Finding out such an approach with stress on insect-plant interactions won't solely facilitate ecological and organic process relations. However, it also helps to develop a unique, novel, and advanced crop protection strategy either by engineering through biotechnology or choosing crop plants that are endogenous for inducing resistance to insect pests, or by treating crops with sustainable pesticides to create less engaging, captivating, and attractive, generating unique methods to draw in males and females of pests and as a means of observation and prediction of populations. With increasing indications on issues relating to indiscriminate use of artificially produced pesticides, it tries to look for different strategies of crop protection and has been a world-wide effort in order to get the best results. Studies on chemical ecology for better use of pesticide chemicals provide such a viable option to get rid of the harmful impact of pests on plants.

1.5.1. Strategies to Enhance Beneficial Plant aspects

One of the foremost powerful and long-lived ways to attenuate economic harm from pests is to spice up populations of already present useful organisms by providing them with applicable surroundings and various food sources. Useful organisms, for example, various predators, host parasites actions and pest-sickening living organisms known as "pathogens" are found more constantly and rapidly on various farms wherever fewer pesticides are used, than in monocultures systems or in field areas where the plants are habitually treated with pesticides. The subsequent characteristics are common in farms where the host population is in abundance and fields are spread within short area and encircled by natural vegetation.

The systems that are used for cropping are numerous and plant populations present around such fields are perennials and flower-bearing plants. Crops are usually controlled and head organically or with smallest controlled agriculture-based chemicals. The soils that are rich in naturally occurring organic matter are also rich in the microbial population contributing to the biological activity are covered with a large amount of vegetation.

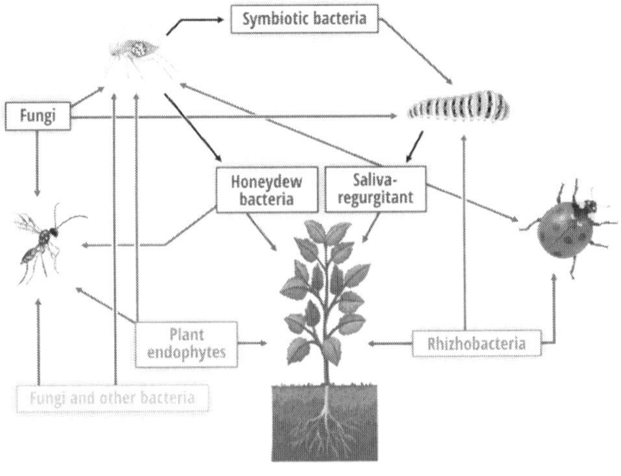

Figure 1.3. Plant–pest interactions mediated by microorganisms.

Source: Image by Wikimedia

To preserve and advance populations of naturally occurring enemies, avoid cropping practices that hurt beneficial ones. Instead, substitute strategies that enhance and promote their survival rates. One should begin by reversing practices that disturb and manipulate the natural biological control; cherish pesticide implementations, hedge degradation, removal and comprehensive weedkiller use are supposed to eliminate weeds in and around the surrounding fields. Even tiny alternations in farming practices can effectively increase and promote natural antagonist populations throughout crucial periods throughout to growing season.

The straightforward use of carefully made straw mulch supply moisture, secure concealment places for nocturnal predators like mostly the species of spiders and ground beetles. By limiting the visual differences between foliage and vacant soil, straw mulch appeared to build it tougher for flying pests most frequently for aphids and leafhoppers to "see" the area or field of the crops they attack. This mixture of effects can considerably scale back insect harm in mulched garden areas or the plots (Altieri, Nicholls and Fritz, 2005).

1.5.2. Strategies to Obtain Multiple Benefits from Plant-Pest Interactions

Researchers from Michigan State University have designed the experiment to evaluate orchard-scale floor cover in commercially available orchards and on newly made species named as tart cherry orchard.

They keenly observed and examined the orchard floors covered with the following ingredients:

- Compost mixture,
- Mulch,
- White clover,
- Crimson clover
- Cerseem clover,
- Crown vetch,
- Indigo vetch,
- alfalfa,
- Rye,
- Annual ryegrass, and
- Buffalo grass.

So far, Results include: There was a rapid growth of recommended mites and the credit goes to the utilization of a pink clover cover crop. Usage of either herbicide or mulch makes a bigger population of pest mites.

Orchards with floor covers irrigated but not handled with herbicides to manipulate the production of weeds appeared to have fruit yields that had been not considerably lower than conventional practices over a half-decade.

Also, irrigation can also be indispensable in this gadget to stop the ground cowl from competing with the fruit trees for water.

Addition of mulch, cover plants or else compost will increase soil natural matter colonies of useful soil microbes and quantities of rich soil inC and N available to trees.

Fewer helpful nematodes, more plant parasitic nematodes and more nitrate leaching had been related to lower-quality conventional-system soils. Poor quality common systems of convention soil are associated with a small number of beneficial nematodes, more destructive plant harming parasitic nematodes and more loss of Nitrogen through nitrate leaching. The study done on straw mulch, should be six to eight inches deep, improved tree boom and yields regardless of the greater population of pest mite.

Nitrate leaching was seen to be at high pace during spring and in fall was extensively reduced by using vegetation growing under timber during these periods.

The Population densities of advisable earthworm, nematodes, and beneficial mycorrhizae associations were greater under a natural production system.

The growth of young trees was seen to get benefits from the addition of mulch or compost; however, the growth can be severely retarded by way of opposition with groundcover flowers as they compete for moisture and essential nutrients.

Trees growing with heavy mulches yield the fruit of soft type in two of seven years. The scientists additionally are inspecting the impact of mixed-species hedgerows on insect pest motion. Also, they are investigating the role of insect pheromone in mating disruption, and the associated mass trapping of plum curculio in 14- inch groundcover bands around mulched center lines.

1.6. ROLE OF NATURAL ENEMIES IN PLANT-PEST INTERACTIONS

Naturally occurring enemies regularly target insect pest species having only single stored product, and more than one species of such enemies may additionally need to be launched when more than single insect pest species is there to compete.

Emancipation of more species of natural enemies usually leads to a considerable increase in the finances of management of insects and pests. Such expenses may also be elevated when a couple of releases of one or greater such enemies are necessarily required.

Sometimes although the timing of the release is perfect, additional pest enemies often might also prove to be needless.

Natural enemies need to be launched while the populations of pests and insects are not large and easily manageable because they might also no longer suppress or retard the population growth of insects and pests. Also, natural enemies of pests do not work as swiftly as chemical methods do.

Quality control is quintessential in the production of pest enemies so that solely the wholesome herbal enemies are offered and the right species of natural enemies are released.

1.6.1. Bark Beetle

Natural enemies, for example, predators, parasitoids, and pathogens, are very attractive mechanisms that play an essential role in the populace dynamics and ecology study of bark beetles in references to ecology and their population dynamics.

Natural enemies and their value in bark beetle population dynamics was a hot topic in European and North American species majorly in coniferous forests. One of the famous natural enemies of bark beetles in coffee where the coffee berry borer *Hypothenemus hampei* is a promising bark beetle pest.

1.6.2 Predators of Bark Beetles

Woodpeckers have been recognized as vital enemies that form an antagonistic relationship with bark beetles that are one of the major problematic concerns in Europe (Sperling, 1878).

A variety of bird groups target bark beetles in their diet, especially catching and consuming them in the time of flight while the beetles are seen

1.7. PLANT-APHID INTERACTIONS

Aphids constitute an important fundamental group of crop pests that restrict productiveness of vegetation and leads to extreme damage to plant life each by using direct feeding and through indirect mean as vectors of many infectious agents. It is a relatively small insect organization (about 5000 recognized species) compared to 10 thousand species of grasshoppers and 12,000 species of geometrid moths; aphids are serious trouble for agriculture. Of the 5000 recognized species in the family Aphididae, 450 are endemic on crop vegetation, and a hundred have effectively exploited the rural environment to the extent that they're of big monetary importance.

The aphids are the specialized phloem sap feeders resulting in large yield losses in lots of plants. It's far their ability to unexpectedly make the most of the ephemeral habitats that make them severe pests and these potential effects are as follow:

- High reproductive rates,
- Efficient Dispersal capabilities, and
- Adaptability survives under unfavorable environment

In comparison to the majority of insects, aphids display parthenogenetic viviparity that is the phenomenon that limits the need for adult males to fertilize females and gets rid of the egg stage from their life cycle.

Therefore, aphids reproduce clonally and produce younger ones, and embryonic development of an aphid starts off evolved earlier. Many of these traits allow aphids to exploit the periods of rapid plant growth time, preserve energy, and permit for brief generation times; mostly certain aphid species can attain maturity in as little as 5 days (Gomes et al., 2013).

1.8. PEST-PLANT INTERACTIONS OCCURRING IN POTATO CROPS

Potato regularly incorporates diverse communities of herbivores and their closely associated natural enemies imparting a perfect model machine to study the ecological troubles related to potato crops.

Traditionally, potatoes have been closely dealt with insecticides, due to the excessive economical value of the crop and the extreme damage brought about by means of insects and different pests (Hare 1990).

Figure 1.5. Potato Crop Health Management.

Source: Image by pakkison.com

However, recent years have seen developing strain to reduce insecticide use on this famous food crop, leading to a growing natural sector and the adoption of pesticide-reduction schemes in conventionally managed fields (Werling and Gratton 2010).

Pest management selections in potatoes may additionally regularly focus on just one or two pests perceived to be of the best financial importance, these crops typically house a diverse community of many herbivore species. Multiple species of herbivores have the achievable to not directly influence each other densities, with these interactions generally transmitted indirectly through the plant host. This approach of having more than one herbivore can be either detrimental or recommended to unique key potato pests. Likewise, the presence of multiple herbivore species can either increase or decrease the influence of naturally occurring enemies on a specific species of pest (e.g., Eubanks and Denno 2000). Species from Herbivore origin can also indirectly influence one any other via altering the behavior of shared predators. For example, prominent competition takes place when the existence of one herbivore species attracts large numbers of enemies than might in any other case be found. When these enemies change to feeding also on another herbivore species, the second herbivore species is harmed (Andow 2004).

One of the vastly studied detailed collections on predation among aphid predators examined interactions amongst predatory midges such as

- Aphidoletesaphidomyza
- Chrysoperla rufilabris
- *C. maculate* attacking potato aphid species (*Macrosiphum euphorbiae*).

The midges are surprisingly inclined to intraguild predation by the lacewing and *C. maculate*. Apparently, in response to its high chance of falling victim to intraguild predation, the midge has developed various strategies to reduce this danger.

First, midge larvae will "hide" at the center of aphid aggregations, so that aphids will soak up most predator attacks. Second, midge eggs are get laid and deposited on plants with high trichome density, which makes them surprisingly immune from predation by *C. maculata* predators (Lucas and Brodeur 1999).

Altogether, this suggests that the selective pressures exerted by way of intraguild predators have led the parasitoid to strengthen its potential to trade host behavior, in order to be delivered to apupation site mostly out of attaining of the parasitoid's own natural enemies (Brodeur and McNeil 1992). Similarly, *C. maculata* leaves aphid-infested potato vegetation to pupate elsewhere, interestingly in order to avoid intra guild predation with the aid of the many predators that aphids entice (Lynch et al., 2013)

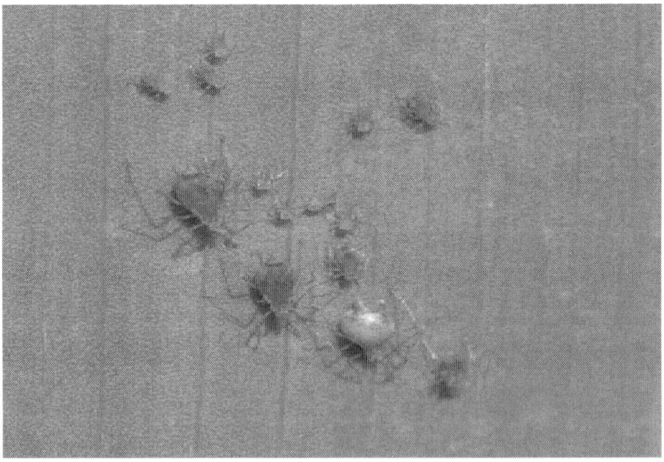

Figure 1.6. Aggregation of aphids with healthy the parasitized individuals.

Source: Image by flickr.com

Increasing herbivore range has in a similar way complex effect. Herbivores such as flea beetles and leafhoppers that cause destruction by their attack early in the growing season start influencing in order to initiate resistance induction in potato flora that renders plants much less alluring or nutritious to later-occurring herbivores like Colorado potato beetles.

As a result, early-season herbivory can dampen later herbivory by means of other species. However, the presence of noticeably captivating prey like aphids may attract predator attacks away from not so attractive prey like potato beetles, such that aphids indirectly protect potato beetles (Lynch et al., 2013) Thus, it is obvious that increasing herbivore diversity can either harm or advantage unique pest species.

1.9. BENEFICIAL PLANT PEST INTERACTIONS

Tree fruit growers looking for options for broad-spectrum pesticides are seeking tomanipulate insect pests by using a greater extra environmentally friendly approach. In Washington state pear orchards, SARE-funded research has discovered that mowing once a month as a substitute than2-3 times a month creates eye-catching habitats for helpful insects.

ARS researcher partly funded through SARE designed specially organized trials at three orchards and different mowing frequencies (weekly, month-to-month and simply once a season).

With less popular mowing techniques, the natural enemies moved into the ground cover in greater numbers, probable attracted to the pollen and nectar newly handy from flowering plants as properly as greater plentiful prey, such as aphids and thrips.

Researcher Dave Horton determined greater lacewing larvae, spiders, ladybug beetles, damsel bugs, parasitoids and minute pirate bugs. “If you mow a lot, you won’t have a lot in the way of herbal enemies on the ground,” Horton said.

“By lowering the frequency to once a month, you see a dramatic make bigger in natural enemies transferring into the ground cover without a massive expands in pests that feed on fruit.”

Questions continue to rise whether-or-not the predators migrate from the floor cover into the pear timber to attack orchard pests, though proof supports that some predators, especially spiders, appeared in greater numbers in pear trees in the less frequently mowed plots, good news for pear growers.

1.10. CREATION OF WAYS FOR NATURAL ENEMIES TO PROMOTE PLANT-PEST INTERACTIONS

One can offer natural enemies with highways of habitat with the aid of sowing various flowering vegetation into strips every a hundred sixty-five to 330 toes (50–a hundred m) across the sector.

Beneficial can use these corridors to flow into and disperse into discipline centers. EU studies have confirmed that this practice will increase the diversity and abundance of herbal enemies.

While sugar beet fields have been drilled with corridors of tansy leaf (*Phacelia tanacetifolia*) each twenty to third rows, destruction of bean aphids by way of syrphids intensified. In addition, strips of buckwheat and tansy leaf in Swiss cabbage fields expanded populations of a small parasitic wasp that attacks the cabbage aphid.

Because of its lengthy summer time flowering duration, tansy leaf has also been used as a pollen supply to enhance syrphid populations in cereal fields. For greater prolonged effects, plant corridors with longer-flowering shrubs.

In northern California, researchers related a riparian forest with the core of a giant monoculture vineyard using a vegetation corridor of almost sixty different plant species. This corridor, which included many woody and herbaceous perennials, bloomed at some stage in the developing season, furnishing naturally occurring enemies with a regular grant of alternative foods and breaking their strict dependence on grape-eating pests.

A complicated of predators entered the vineyard sooner, circulating constantly and completely through the vines. The subsequent food-chain interactions enriched populations of herbal enemies and curbed numbers of leafhoppers and thrips. These influences have been measured on vines as a long way as 100-150 ft from the corridor (Altieri, Nicholls and Fritz, 2005).

1.11. EVOLUTION AND PLANT-PEST INTERACTIONS

The evolution of vegetation transformed the terrestrial surroundings into a quite treasured aid for the herbivore community. In herbal ecosystems, vegetation and insects are simply some of the dwelling organisms which might be constantly interacting in a complex way.

Those organisms are in detail associated when you consider that insects have several beneficial activities, which include protection and pollination while vegetation provide habitats, ovi-position sites and meals, the three principal elements asked for insect proliferation (Panda and Khush, 1995). On the other hand, depending on the depth of insect attack, herbivores might be extraordinarily dangerous to plants as a loss of life.

Plants throughout their life produce chemical substances for protection purposes in two special ways; first, as constitutive materials to repel herbivores through direct toxicity or by reducing the digestibility of plant tissues and second, as inducible materials synthesized in response to tissue harm through herbivores. Those techniques can prevent maximum of the herbivores although there is a reduced range of insects that might be capable of adapt to specific plant species.

Secondary metabolites perform useful features for the plant acting both in an inducible or constitutive manner. Some compounds are plant increasing regulators even as others act as chemical indicators in the atmosphere, antibiosis agents, transport and storage of carbon and nitrogen molecules, which might be directly involved inside the plant primary metabolism. Secondary plant compounds are concerned in plant defense towards insect herbivores appearing as insect repellents, feeding inhibitors and/or pollution.

Compounds are any other important mechanism affecting the conduct of insects attempting to find food. Specific smells or odors are usually released from flowers are one of the number one cues that insects use to discover the host plant. As an example, cabbage seed weevil appears to be orientated through a complicated mixture of host plant volatiles.

The presence of chemical volatile compounds in vegetation can act in two ways. First, they can repel an extensive range of potential herbivores due to the character of toxic compounds released inside the air.

The second way is their property of attracting a small range of specialized pest species and of appearing as an indirect plant protection mechanism by attracting other bugs that prey on or parasitize the herbivores (Birkettet al., 2000).

The volatile compounds are released in the tobacco plant draws predatory bugs to tobacco hornworm eggs and feeding larvae dramatically increasing the predation costs. Moreover, those secreted volatiles decrease oviposition prices from grownup moths because adults avoid plants on which predators are likely to be present, decreasing herbivore loads to approximately 90%. Plant volatiles can also act as signals between plants, in which volatiles from a broken tissue induce protection response in neighboring undamaged flora (Fleurat-Lessard, 2015).

Further to a local inducible synthesis of PIs, it was proven that indicators from the broken tissue are transported through phloem and stimulate the synthesis of PIs at some stage in the plant (Jongsma and Bolter, 1997).

Plant proteinase inhibitors characteristic as specific substrates for the digestive proteinases, forming a stable complicated wherein proteolysis is restrained and extraordinarily slow (Tiffin and Gaut, 2001).

In the end, proteinase inhibitors act by means of inducing an amino acid deficiency influencing the insect boom, improvement and ultimately inflicting their dying both by inhibition of intestine proteinases or because of a huge overproduction of the digestive enzymes, decreasing the availability of vital amino acids for the manufacturing of other proteins.

Slowing herbivore increase, those inhibitors prolong the time that predators will be attracted to plant life via the volatile release. Direct defenses that amplify the time for the duration of which larvae remain in those instars would likely increase the effectiveness of the indirect defense (Baldwin 2001).

Transgenic tomato flora over-expressing the ProSystem in transgene has been shown to build up a soluble cystatin constitutively. In leaves of wild-type tomato plants, cystatin mRNA amassed systemically in response to wounding, remedy with methyl jasmonate and whilst supplied with systemin (Siqueira-Junior et al., 2002).

Moura and Ryan (2001) tested that the induction of proteinase inhibitors in pepper in response to herbivores is regulated through the octadecanoid signaling pathway. But there are some variations inside the manner in pepper while as compared to the system formerly defined for tomato (Pearce et al., 1991).

According to the authors, this may imply that PIen coding genes in exceptional species may have evolved to protect them in opposition to precise pests and pathogens belonging to their specific ecological niches.

Except for a rapid response to insect attack, prompted PIs can play a preventive function in plant protection. As an instance, attacked *Populus tremuloides* leaves accumulate trypsin inhibitor within 2 days in turn to reduce subsequent damage (Haruta et al., 2001).

During evolution, flora and bugs developed ecological, physiological and biochemical mechanisms

to weaken the impact of insect proteinases and plant proteinase inhibitors, respectively. Vegetation evolved amazing traits towards insect proteinases via increasing the inhibitor pastime of their tissues (Rakwal et al., 2001) This can entail developing a heterogeneous set of inhibitors owning activities to diverse enzymes, by way of producing bi-functional inhibitors which can be lively to amylases and proteinases, by way of growing the complexity of inhibitors with distinct biochemical properties through the production of isoinhibitors.

When an insect species develops a mechanism that permits it to apply a toxic plant for food, numerous beneficial are located. This plant can be without difficulty diagnosed through its secondary compounds and constitutes a supply of food that may be used by very few different herbivores, minimizing the opposition.

This has a shielding implication because the antibiotic residences of many poisonous chemical compounds might also protect the herbivore towards pathogens and feeding in these might also impart a toxic or unpalatable feature to the herbivore, that in turn reduce predation (Fleurat-Lessard, 2015).

1.14. EVOLUTIONARY AND ECOLOGICAL PROSPECTIVE OF PLANT-PEST INTERACTIONS

Most creatures don't really exist in a uniform world, but rather in one that is constantly changing in both space and time. The complex, diverse chemical environments created by both host and non-host plants affect the interactions between plants, herbivores, and their natural enemies. When under selection pressure, insect pests and their natural antagonists must deal with the temporal and geographical variability of host plants.

This sensory system and its search processes should have evolved to maximize their fitness by effectively identifying suitable host plants. A complete understanding of evolution in wild populations necessitates an approach that incorporates interactions between organisms and the environment. It is feasible to explain the genesis and preservation of patterns of variation at multiple hierarchical levels.

In practice, this encompasses the examination of the genetic background of different phenotypes, how these phenotypes perform under different environmental conditions, and how they are constrained by significant biological trade-offs.

In plants, mutualisms and antagonists work together to shape various phenotypes. Non-pollinator agents, such as herbivores, are increasingly being recognized. Plant-herbivore interactions are one of the most common interactions in nature and have significant evolutionary implications as herbivores devour around 10-15% of plant biomass generated each year. Herbivores may have a dramatic effect on plant structure and composition, influencing the shape of natural ecosystems.

At the heart of a thriving community of scientists is the study of insect–plant interactions, which encompasses a broad variety of biological problems ranging from the molecular to the ecosystem level, all of which are linked by evolutionary biology.

The relevance of interdisciplinary study in understanding the intricate connections among plants and insects cannot be overstated. In recent years, new technologies and analytical techniques, including next-generation sequencing (NGS) and gene-editing technology, have radically altered this field of study (e.g., CRISPR-Cas9).

There have also been advances in *in vivo* imaging and greater chemical analysis, among other areas. Molecular biology, genomics, chemistry, physiology, and other techniques may now be carried out under controlled laboratory circumstances as well as in natural environments, with the results being integrated.

These advances should allow us to gain a more comprehensive understanding of complex ecological networks, the physiological, ecological, and evolutionary dynamics of these encounters, as well as the genetic basis of traits, as well as to analyze data that was earlier impossible to answer by conventional methods.

It is necessary to comprehend insect and microbial effectors mostly on one hand, as well as plant responses on the other, in order to decipher the molecular processes underpinning insect–plant interactions. Plants, on the other hand, must interact with a diverse range of biotic partners, extending from parasites to mutualisms.

It is inevitable that the signaling networks that are triggered by plants respond to parasitic, herbivorous, and helpful species will overlap in their functions. The management of the evolutionarily conserved response of the plant must strike a delicate balance between defense and the development of beneficial properties (Endara et al., 2017).

However, the capacity to receive and modify plant signals confers unexpected adaptive skills on insects or diseases, allowing them, for example, to spread into new ecological niches (Fordyce, 2010). Increasing our awareness of insect–plant relationships involve not only a greater understanding of the cellular and molecular conversation that occurs between plants and insects, but also the investigation of the ecological and evolutionary consequences that result from these interactions.

Researchers from the fields of molecular biologists, geneticists, ecologists, and evolutionary biologists are working together to achieve this aim and integrate the biological foundation of insect–plant interactions into an ecological framework.

It is possible to get new insights into the intricacy of induced signaling networks that have developed between plants and their predators through the study of multitrophic interactions. It is possible to identify adaptations that enabled ecological diversification, specialization, and speciation by combining knowledge of molecular processes of insect–plant interactions with phylogenetics and evolutionary genomics.

Ultimately, understanding how plants fight themselves against phytophagous (or herbivore) insects, as well as how insects adapt to their environment, would be beneficial to agricultural and forestry research and development.

These agricultural systems, it is becoming increasingly evident, rely on the ecosystem services supplied by natural ecosystems to function properly (Rusch et al., 2017). Since natural and agricultural ecosystems are facing significant environmental problems, such as those caused by climate change, the most recent study can assist in better understanding and mitigating the environmental crisis that we are now experiencing.

Recent advances in thermal ecology are showing how ecological relationships will be affected by global climate change, as well as how humans might be able to prevent their negative consequences.

1.15. INSECT EFFECTORS AND PLANT RESPONSES

Plants' response to insect herbivores has an impact on the plant-dwelling players as well as the plants' relationships with other creatures connected with the plants (Franco et al., 2017). Plants use both DAMPs and HAMPs, also known as elicitors, to detect herbivores. Following detection of DAMP and HAMP, various defense mechanisms are engaged to reduce the herbivores' harm (deterrence).

Plants have developed the potential to limit and control these reactions, including effectors which are injected into the plant (Kaloshian & Walling,

2016). Although effects on herbivore performance have been reported in studies on aphids and spider mites, effects on plant defensesignaling and responses have not been detected in these studies.

It is widely believed that nutrient-absorbing herbivores, such as galling insects, may transfer growth hormones into their plant hosts, which promote nutrient depletion and exhibit many varied morphologies. With current advancements in the area, the research has made it possible to identify insect and plant elicitors and discover the details of the induced plant responses.

This large-scale genomic sequence analysis, transcriptomics, proteomics, investigational evolution, and genomic sequence connection holds great promise to expand our knowledge of how various herbivores (i.e., non-model and model) interact with different plant hosts (i.e., model and non-model) to understand how such patterns of genomic evolution (such as gene expansions) relate to insect range and feeding strategy, comparative genomics of generalist and specialized insects might be useful.

While quantitative genetics is also needed to understand better the genomic architecture underpinning host changes, specialization, and journalism in insects, a large amount of research should also be done for the sake of this investigation.

The outcomes of this research will influence the success of future gene discovery and functional validation by identifying potential gene interactions between loci controlling insect traits for plant specialization, and characterizing the genomic regions under the QTLs for further genetic analysis and functional validation (Kanvil et al., 2015).

Over the last decade, several genes and their roles that are believed to be involved in host range evolution have been discovered, and it is quite likely that more will be found soon. There is currently no functional information to interpret the precise role of a gene in its interaction with the host plant. In this research, gene ontology analysis is done using gene prediction databases to identify undiscovered genes. In order to demonstrate their involvement and importance in host plant selection and exploitation, functional studies of these genes are required. Using several technologies, including transcriptomics, proteomics, and metabolomics, is an excellent strategy for identifying genes that are candidates for functional analysis. This, in conjunction with evolving molecular techniques like CRISPR-Cas9 genome editing, enables functional validation of gene candidates in non-model species (e.g., yeast) (Koutroumpa et al., 2016). Also, with regards to plant defenses against insects, there is an increase in the number of plant systems being investigated (Züst & Agrawal, 2016). By expanding the understanding of insect-plant relationships, it may be possible to better understand the evolutionary histories of the battle between adversaries and its implications for diversification patterns.

There are new targets for managing crop pest insects as a result of the breakthrough in insect–plant communication. Selecting plants that are less susceptible to salivary proteins can be achieved via knowledge of saliva protein activities. Using these chemosensory genes, one may develop attractant and repellent plant volatiles to maximize the effectiveness of attract-push techniques (Zhang et al., 2017).

This facilitates the accumulation of more strong data on the impact played by plant specialization on insect diversification and speciation. In this way, the process of genetic differentiation accumulation will be more fully comprehended, and enhance our knowledge of how genetic diversity is evolved when species differentiate (Nosil et al., 2017).

One form of plant defense against pathogen infection is triggered by the ingestion of plant tissue by an herbivorous insect. This organism hosts microbial communities and, as a result, it may broaden the insect's opportunity to manipulate plant species and regulate plant primary and secondary metabolisms and/or defenses against parasites. It is thus essential to comprehensively account for the diversity, dynamics, and roles of microbial communities associated with insects and their host plants when using evolutionary genomics of insect–plant interactions.

1.16. PLANT RESPONSES TO INSECT ATTACK

Various plants use different strategies to generate chemicals for defense. Constitutive compounds deter herbivores by direct toxicity, whereas inducible chemicals are produced in reaction to harm from herbivores. The techniques that have been used here are able to keep most herbivores under control, but there are fewer insects that can shift to different plant species. Some secondary metabolites help the plant, either inducing or constitutively, to do their intended task. Plant growth regulators and chemical signals all serve similar roles in the environment, whereas antibiosis agents and carbon and nitrogen molecules all help to promote plant development.

Plant secondary chemicals help deter insect herbivores, impede growth, and/or serve as poisons. Using this research, we divide these hazardous chemicals into those that are chemically derived and those that are produced from proteins. The second-order metabolites like glucosinolates are an example of how plants and insects affect each other. There are several qualitative classes of molecules in Arabidopsis plants, and the different combinations they generate in response to herbivory or other selected pressures depend on the type of selective pressure present. Another significant method regulating insect behavior is the release of volatile chemicals. Insects rely on plant odors to find the host plant. Cabbage seed weevil seems to use a complicated combination of plant volatiles to find food.

According to this, plants contain volatile chemical molecules, implying a twofold meaning. Due to the toxicity of poisonous chemicals emitted in the air, these plants can repel a wide spectrum of possible herbivores. Secondly, they are selective in what animals they lure. They also function as a trap for other insects, which feast on or parasitize the herbivores. The volatile mixture of the tobacco plant draws insect predators to the eggs and larvae of tobacco hornworm, leading to an exceptionally high level of predation, as reported by Kessler and Baldwin (2001). The release of volatile chemicals also reduces egg laying in adult moths because adult moths steer clear of plants where predators are expected to be present, cutting moth populations by 90%. Volatile chemicals produced by plants can serve as signals between plants, where injured tissue emits detectable volatile molecules to alert undamaging plants of injury. Each plant releases its own herbivore-specific blend of volatile compounds when a particular herbivore species feeds on the leaves. This implies that there is a co-evolutionary relationship between herbivores, natural enemies, and plants.

Wounds caused by herbivory provide an opportunity for opportunistic microbes to establish and provide a pathogen risk. However, the induced resistance permits the plant to reach its full capacity for either herbivory or disease tolerance. The fact that insect-eating triggers the synthesis of phytoalexins, which have antibiotic capabilities, has been demonstrated by research. When intact cells surround injured tissue, the cells' cellular membranes strengthen and close the wound. In addition, intact cells around areas of damaged tissue physically barricade pathogens by strengthening the cell wall and closing the wound.

Populus tremuloides uses a wide range of defense mechanisms that have both protein and phytochemical-based components against insect herbivores. Most plant-herbivore interactions have defensive systems that are sophisticated. While several plant chemicals are excellent in reducing insect damage, they are highly expensive in most cases. Tissue damage and significant reserves necessary for resistance both occur during an herbivore assault. Herbivory decrease and fitness will move in tandem as resistance evolves. Thus, plants may be able to avoid costs associated with allocation by developing alternative metabolic processes that are more efficient. The usage of inducible defenses (sometimes referred to as "plant's expenditure minimizers") provides a less costly option when it comes to making investments in plant defense, as well as how to react to non-browsing herbivores. In certain cases, some plants employ repression of the early and expensive defensive events rather than resistance as their defense mechanism (Baldwin, 2001).

1.17. WOUNDING SIGNALING PATHWAYS

Ion fluxes from across plasma membrane, variations in cytoplasmic concentration, active oxygen species production, and protein phosphorylation patterns alter after the initial event has occurred. During wound healing, these occurrences result in the creation of cell wall fortification at the affected area, metabolic changes, and the production of signals, which influence the expression of defense genes.

Mechanical wounding is a component of plant damage, although mechanical and thermal wounding are two distinct processes. Consider the following scenarios: Herbivore eating and larval oral fluids or regurgitate as well as the production of volatiles that attracts parasitoids operate in concert to raise wound-induced JA levels.

Because some of the variations may be due to distinct feeding mechanisms, while others might be due to different elicitors, different feeding mechanisms and different elicitors could both be implicated. Herbivores produce chemicals in their mouth secretions. a low-molecular-weight fatty acid derivative, N-(17-hydroxy-inolenoyl)-L-Gln (volicitin), has been found in caterpillars of the beet armyworm (Alborn et al., 1997).

When plant leaves are damaged and then expose to the elicitors, numerous previously unknown, poorly described events occur. The octadecanoid signaling cascade is activated during tissue injury, which causes the release of systemin into the circulatory system. As described, several reactions lead to JA biosynthesis. The synthesis of signal pathway genes in the vascular bundles is also increased, as well as H2O2, which is a second messenger that activates putative defense genes (late genes) in mesophyll cells, such as the antifeedant proteinase inhibitor genes.

Even though JA is the most prevalent defensive signal for chewing insects, research has shown that ethylene is a significant modulator of plant defense in several species, working concurrently or sequentially with JA in the leaves of the plants that sense the insects (Stotz et al., 2000).

1.18. PLANT DEFENSE ASSOCIATED WITH STORAGE PROTEIN

While many of these proteins possess entomotoxic characteristics such as amylase and proteinase inhibitors, lectins, and globulins, storage proteins synthesized and accumulated in plants are known to be strongly associated with plant defense. Most leguminous plants, including beans, include these proteins in seeds and vegetative organs.

When plants are attacked by insects, experience mechanical injury, or are subjected to exogenous phytohormones, proteinase inhibitor (PI) levels can be highly elevated in the leaves. Additionally, it was shown that injured tissue transmitted phloem signals induce the creation of PIs inside the plant. Inhibiting the digesting enzyme by combining with it to create a stable complex that inhibits proteolysis; proteolysis is greatly impeded and slowed (Tiffin and Gaut, 2001).

Proteinase inhibitors help by producing an amino acid shortage, which has shown to negatively influence insect growth, development, and ultimate mortality owing to either intestinal proteinase inhibition or excessive digestive enzyme synthesis (Jongsma and Bolter, 1997; Pompermayer et al., 2001). Inhibitors that restrict the growth of herbivores delay the time predators might be drawn to plants, prolonging how long predators will have to hunt. While larvae stay in these instars, direct defenses will have a greater impact (Baldwin 2001).

Thus, transgenic tomato plants overexpressing the ProSystem in transgene consistently accumulate a soluble cystatin, even if they do not have elevated levels of the endogenous enzyme. At the time of wounding, systemic accumulation of cystatin mRNA occurred in wild-type tomato plants in response to treatment with methyl jasmonate, and with systemin supply.

Results show that tomato cystatin prevents the proteolytic activity of midgut homogenate from third-instar larvae of *Callosobruchus maculatus* and *Zabrotes subfasciatus*. Pepper herbivore defense induction is under the control of the octadecanoid signaling system. While on the other hand, there are significant distinctions in the process of peppers, as opposed to the one for tomatoes.

It is posited that different species' PI-encoding genes may have evolved to protect their habitats from distinct pests and diseases. Another way in which induced PIs help with plant defense is in speeding up the plant's reaction to insect assault. For example, damaged *Populus tremuloides* leaves develop a trypsin inhibitor over the course of two days, which is intended to minimize the amount of damage following the first assault.

Different species acquired a variety of physiological, biochemical, and ecological strategies to hinder insect and plant proteinases over evolution. Entomophagous organisms have discovered several strategies to prevent themselves from suffering the ill effects produced by the proteinase inhibitors present on their plant hosts. Changes include: increases in the activity of digestive enzymes, decreases in the activity of digestive enzymes, changes in the spectrum, or alterations in the relative activities of various digestive hydrolases.

Lectins are carbohydrate-binding proteins, which are predominantly found in legume plants, and are mostly located in the plant's storage organs and structures that protect it (Ramos et al., 2001). Phytohemagglutinins, arcelins, and a-amylase inhibitors make up the bean-lectin gene family. Lectins found in plants have been investigated for their insecticidal effects on different insect orders.

The jack bean lectin, found to affect insect development in both the Lepidoptera (Laoleracea) and Hemiptera (*M. persicae* and *R. prolixus*) orders, was shown to impede the process in at least two insect orders, the Lepidoptera and Hemiptera (Ferreira-da-Silva et al., 2000).

Leguminous seeds include various classes of storage proteins in the globulin family, which are called vicilins. These products bind tightly to chitin-containing structures present in the midguts of insects and yeast and filamentous fungus, making it difficult for the invading organism to grow and expand (Sales et al., 2000).

The proteins present in cowpea (*V. unguiculata*) seeds protect the seeds from the larval stages of *C. maculates*. Since *C. maculatus*'s midgut enzymes only hydrolyze vicilins at a low rate, some of this resistance is attributable to it (Sales et al., 1992).

colonization. Due to the process of evolution, herbivores specialized to feed on certain plants have developed unique abilities to utilize plant-based compounds as mating attractants. Plant defense chemicals are often detoxified or sequestered by these insects, and they can cause parasites and predators to be unappealing or poisonous.

Now sequestration experts can take these chemicals and deposit them in the larvae, the adult body, and even in the eggs with almost complete impunity (Nishida, 2002). Additionally, these chemicals help supply insects with signals that aid in host identification, enabling fast and efficient host-finding activities during feeding and oviposition.

Even if secondary metabolites of a non-host plant can discourage specialists that are equally sensitive to these phytochemicals, on the other hand, primary metabolites of a non-host plant may also act as deterrents to generalists that also exhibit a similar level of sensitivity. The results indicate that experts are more likely to select better hosts in the face of combined hosts (Bernays, 2001).

Since highly valuable plant species are now being put together in research communities, Collaborations often form between participants in a genomic project. It encourages multidisciplinary dialogue and the exchange of knowledge and technology by creating a community for plants that continues beyond the release of the genome.

This is vital because plant scientists working on interactions between insects and plants should be members of these networks of plant scientists. When including arthropods in the process of constructing ideal plants, one must consider ways to protect against massive numbers of harmful pathogens, increase pollinator and beneficial symbiont presence, and make use of nutrients and water more efficiently.

It's very probable that disputes will emerge when the ideal plant is formed. For example, a conflict between 'perfecting' the plant for protection against herbivores vs. diseases, or a conflict between 'perfecting' the plant for protection against herbivores vs. promoting pollinators may both be relevant in cases where biotic interactions are concerned.

The Wheat Initiative (wheatinitiative.org) represents a research community that has joined together around a valued plant species to coordinate worldwide research on wheat. A recently established Expert Working Group on Pests and Pathogens ("Control of Wheat Pests and Pathogens") recently published an overview of the diverse wheat pests (arthropods, nematodes, and rodents), new pest management technologies, the impact of climate change on wheat insect pests, and case studies in New Zealand and Australia (Horrocks et al., 2017).

The world's top plant breeders choose the characteristics that will be added to elite cultivars. Only when they are aware of the trait and the evidence that it yields does this selection process take place. Plants might be engineered to contain traits that impart both direct and indirect resistance. For example, resistance genes and toxins are both traits engineered into plants, but these other traits that improve pest control are additional traits that may also be included. Groups of cabbage plants have different levels of their inherent enemy attraction abilities. In order to cause aphids to move away from the wheat and to draw in parasitoids, wheat was bred to generate an aphid alarm pheromone (Bruce et al., 2015). Among the most efficient, inexpensive and sustainable means of pest control is the development of organisms with increased resistance or tolerance to phytophagous insects.

The plant-insect co-evolution is very interesting. While insects have evolved many strategies to dodge the host plant's protective systems, plants have discovered effective methods to defend themselves against herbivory. Plant-insect interaction has been investigated by numerous workers, although our understanding of the relationship is still minimal. In order to be able to use new, more effective ways for insect pest management, we must better understand this process so we may create new plant types with more-resilient chemical defenses.

1.20. CONCLUSION

Plants constitute a major part of the ecosystem. Therefore, it is inevitable for them to interact with the rest of the constituents that make up the ecosystem. If the observation is done for the agricultural ecosystem, the crop plants form the major chunk of the ecosystem, and thereby, are in a perennial interaction with the rest of the biotic and abiotic factors of the ecosystem. The chapter thus concludes and aims at the interaction amongst the two major components within the agricultural ecosystem; namely, crop plants and insect pests.

Series, Vol. 81: Insect-Plant Interactions in a Crop Protection Perspective (ed. by N Sauvion, P-A Calatayud & D Thiéry), pp. 333–360. Elsevier, London, UK.

27. Kaloshian I & Walling LL (2016) Hemipteran and dipteran pests: effectors and plant host immune regulators. Journal of Integrative Plant Biology 58: 350– 361.
28. Nosil P, Feder JL, Flaxman SM & Gompert Z (2017) Tipping points in the dynamics of speciation. Nature Ecology and Evolution 1: 0001.
29. Zhang R, Wang B, Grossi G, Falabella P, Liu Y et al. (2017) Molecular basis of alarm pheromone detection in aphids. Current Biology 27: 55–61.
30. Züst T & Agrawal AA (2016) Mechanisms and evolution of plant resistance to aphids. Nature Plants 2: 15206.
31. Koutroumpa FA, Monsempes C, François M-C, de Cian A, Royer C et al. (2016) Heritable genome editing with CRISPR/Cas9 induces anosmia in a crop pest moth. Scientific Reports 6: 29620.
32. Kanvil S, Collins CM, Powell G & Turnbull CGN (2015) Cryptic virulence and avirulence alleles revealed by controlled sexual recombination in pea aphids. Genetics 199: 581– 593.
33. Baldwin IT (2001).An ecologically motivated analysis of plant-herbivore interactions in native tobacco. Plant Physiol. 127:1449-1458.
34. Jongsma MA, Bolter C (1997) The adaptation of insects to plant protease inhibitors. J. Insect Physiol. 43:885-895.
35. Ferreira-da-Silva CT, Gombarovits ME, Masuda H, Oliveira CM, Carlini CR (2000) Proteolytic activation of canatoxin, a plant toxic protein, by insect cathepsin-like enzymes. Arch. Insect Biochem. Physiol. 44:162-171.
36. Tiffin P, Gaut BS (2001) Molecular evolution of the wound-induced serine protease inhibitor wip1 in Zea and related genera. Mol. Biol. Evol. 18:2092-2101.
37. Rakwal R, Agrawal GK, Jwa NS (2001) Characterization of a rice (*Oryza sativa* L.) Bowman-Birk proteinase inhibitor:tightly light regulated induction in response to cut, jasmonic acid, ethylene and protein phosphatase 2A inhibitors. Gene 263:189-198.
38. Ramos MV, Sampaio AH, Cavada BS, Calvete JJ, Grangeiro TB, Debray H (2001) Characterization of the sugar-binding specificity of

the toxic lectins isolated from *Abrus pulchellus* seeds. Glycoconj. J. 18:391-400.

39. Sales MP, Macedo MLR, Xavier-Filho J (1992) Digestibility of cowpea (*Vigna unguiculata*) vicilins by pepsin, papain and bruchid midgut proteinases. Comp. Biochem. Physiol. 103B:945-950.
40. Sales MP, Gerhardt IR, Grossi-de-Sá MF, Xavier-Filho J (2000) Do legume storage proteins play a role in defending seeds against bruchids? Plant Physiol. 124:515-522.
41. Panda N, Khush GS (1995) Host plant resistance to insects. CAB International, Wallingford. Bruce et al., 2015).
42. Price PW (1997) Insect Ecology. 3rd edn. John Wiley & Sons, Inc, New York.
43. Berenbaum MR, Zangerl AR (1998) Chemical phenotype matching between a plant and its insect herbivore. Proc. Natl. Acad. Sci. USA 95:13743-13748.
44. Scott JG, Wen ZM (2001) Cytochromes P450 of insects: the tip of the iceberg. Pest Manag. Sci. 57:958-967.
45. Nishida R (2002) Sequestration of defensive substances from plants by lepidoptera. Annu. Rev. Entomol. 47:57-92.

Chapter 2

Direct and Indirect Mechanism for Inducing Resistance to Plants against Disease Causing Insects

CONTENTS

Recent advancements in science and technology have opened the doors for a variety of techniques that lead to a better understanding of the overall behavior of plants and the way they respond to different kinds of pests under different environmental conditions in order to produce better quality crops to fulfill our needs. Pests refer to any harmful insects or animals that can cause massive destruction by attacking plants, animals, forests and even humans.

2.1. INTRODUCTION

Insect pests are one of the major reasons that cause massive world agricultural destruction of food crops and lead to considerable financial and environmental cost loss. Crops naturally hold intrinsic mechanisms for their defenses to guard themselves against such pests, including an extensive array of specialized secondary metabolite-based defense chemicals.

These chemical compounds can be triggered upon strike as in phytoalexins or constitutive in nature for example in the form of phytoanticipins, and can have an immediate influence on the pests or be used in a roundabout way to attract their natural enemies. The structure part of global fingers race between the plants and their insect pests, with the insects creating techniques to guard themselves through the following methods:

- Suppression
- Avoidance,
- Detoxification, or
- Plant seizing of their hosts protecting chemicals.

Utilization and enhancement of the chemical defense competencies of crops have the manageable possibility to resource in the continuing struggle to promote or enhance agricultural pest control and management. Such techniques encompass breeding for the restoration of defense chemical compounds from ancestral varieties, or can go ahead with cross-species transfer of defense metabolite synthesis and regulation (Yactayo-Chang et al., 2020).

Plant defenses in opposition to insect herbivores can be divided into two following types:

- Static Defense
- Constitutive Defenses
- Active or Prompted Defenses

Even though the insecticidal compounds or proteins involved are frequently the equal. Induced defenses have components common to all vegetation, whereas the build-up of constitutive defenses is species-particular.

Insect herbivores spark off induced defenses each regionally and systemically by way of signaling pathways involving the following important chemicals:

- Oligogalacturonic
- Systemin
- Hydrogen peroxide
- Jasmonate

Vegetation additionally responds to insect damage and attack by way of producing volatiles, which may be used to discourage herbivores, to communicate between elements of the plant, or among vegetation, to result in defense responses.

Plant volatiles are also a crucial element in indirect defense. Herbivorous bugs have adapted to tolerate plant defenses.

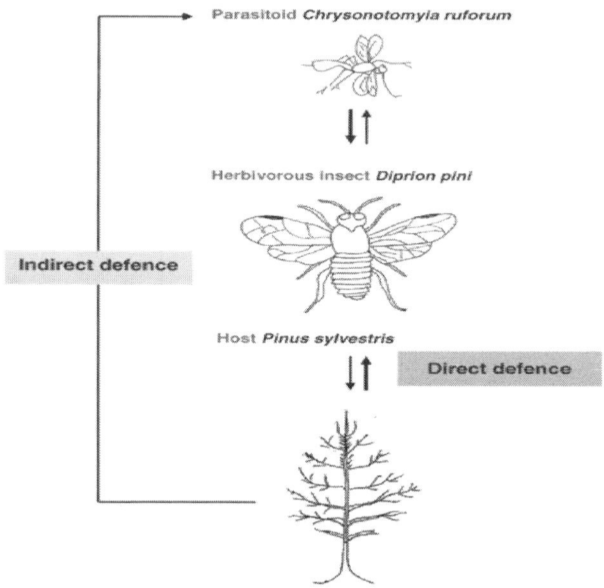

Figure2.1. Direct defense by directly harming the pest and Indirect Defense in *Pinus sylvestris* by attracting natural enemies.

Source: Image by sciencedirect.com

And such diversifications can also be constitutive or induced. Insects whose plant host range is restrained are much more likely to reveal constitutive variation to the insecticidal compounds they will stumble upon, while bugs that feed on a variety of plant species often use prompted adaptations to conquer plant defenses.

Both plant defense and insect versions involve a metabolic fee, and in a natural gadget maximum pant – insect interactions concerning herbivory reach a 'stand – off' where both host and herbivore live on but develop insignificantly.

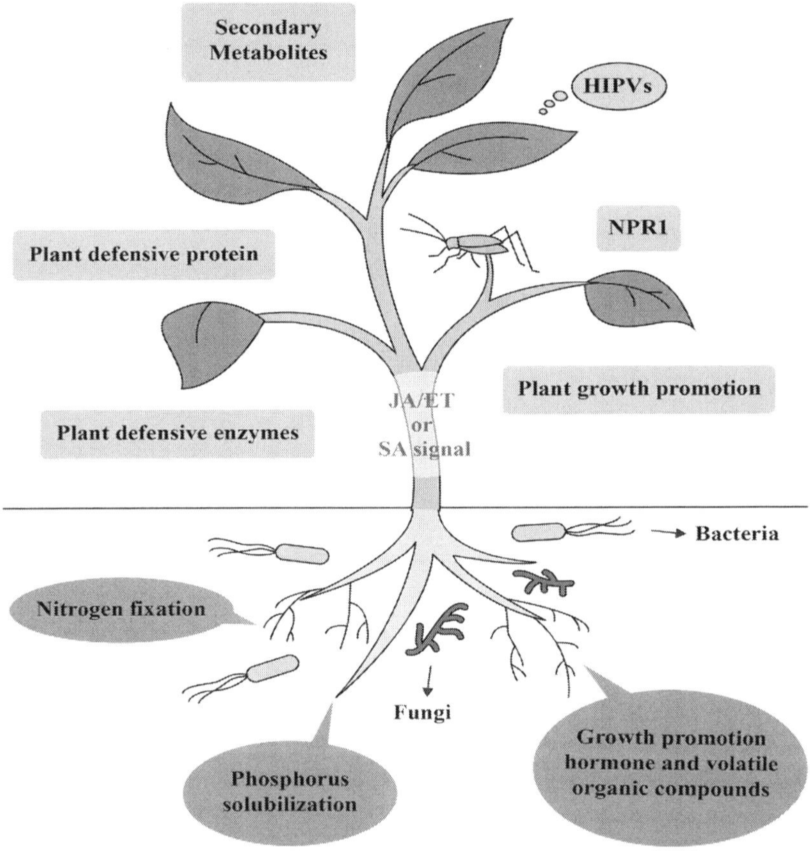

Figure 2.2. Diagramatic Representation of how a plant respond to insect attack.

Source: Image by Wikimedia

To continue to exist and reproduce, flora ought to be able to defend themselves from destructive attacks of pests. Such attacks come from a variety of organisms consisting of microbes and animals; however, several of the most important consumers of flora are insect herbivores.

Many recognized insect species are phytophagous that feed on various parts of flowers other than pollen and nectar, and insect pests' purpose large worldwide crop losses both by way of direct damage and by way of the transmission of plant diseases. Plant life has a variety of inducible and constitutive protection machinery through which they shield themselves from adverse pest attacks. These consist of structural defenses which include thorns and waxy cuticles along with protein-assisted and chemical-based defenses (Pedras et al., 2015).

One of the plant-based defense chemical substances is the production of secondary metabolites. The centers of the structures of secondary metabolite are majorly made of the following components:

- Terpenes
- phenylpropanoids
- Flavonoids
- Nitrogen-containing compounds
- Benzenoids

The identity and purposeful characterization of plant protection compounds is on-going with a focus on compounds determined in important crops and types of structures, or compounds with acknowledged medicinal properties.

But many compounds haven't begun to be structurally confirmed, and others have restrained statistics approximately their production or function. Plant defense chemicals are mostly classified and grouped by using if are secreted by inducible effect or may be constitutively synthesized within the plant. The production of defensive chemicals from the plants has been characterized based totally on their response to microbial pathogens and were named as phytoalexins and phytoanticipins, respectively.

2.1.1. Phytoalexins & Phytoanticipins Definition

Phytoalexins are defined as "low molecular weight, antimicrobial compounds which are each synthesized by means of and accumulated in vegetation after exposure to microorganisms, and phytoanticipins "low molecular weight antimicrobial compounds present in plants" earlier with the aid of microorganisms or produced after contamination completely from pre-existing agents within the plant body (VanEtten, 1994). As lots of those protection compounds are powerful in opposition to more than one variety of organisms, those definitions have on account been broadened to encompass molecules that respond to numerous biological hazards along with insect pests.

The protective mechanisms in plant species operate at ranges. They range from outside defenses like thorns to complicated chemical responses to the poisoning of the attacker. To conquer the insect attack, flowers produce specialized morphological structures or secondary metabolites and proteins that have poisonous, repellent, and/or anti-nutritional results at the insect pests.

Similarly, plant life also launches volatile organic compounds that entice the natural enemies of the herbivores (Pegadaraju et al., 2005).

Resistance elements for direct plant protection towards herbivores insects comprise plant tendencies that negatively influence insect preference through the selection of host by the pest, the egg depositing through oviposition and the feeding behavior or overall performance resulting in improved plant health in an opposed environment.

Such developments encompass morphological functions for bodily defense, like :

- Spines
- Thorns
- Wax crystals
- Epicuticular wax
- Tissue durability

In addition, they consist of compounds for chemical protection, like secondary metabolites, digestibility lowering proteins, and anti-nutritive enzymes.

Most of these traits may be expressed constitutively as preformed resistance factors, or they may be inducible and deployed only after attack by way of insect herbivores.

The induction of protecting developments is not constrained to the web site of assault but extends to non-infested wholesome parts of the flora.

The systemic nature of plant responses to herbivore assault necessitates a protracted-distance signaling system capable of generating, transporting, and interpreting alarm signals produced on the plant–herbivore interface.

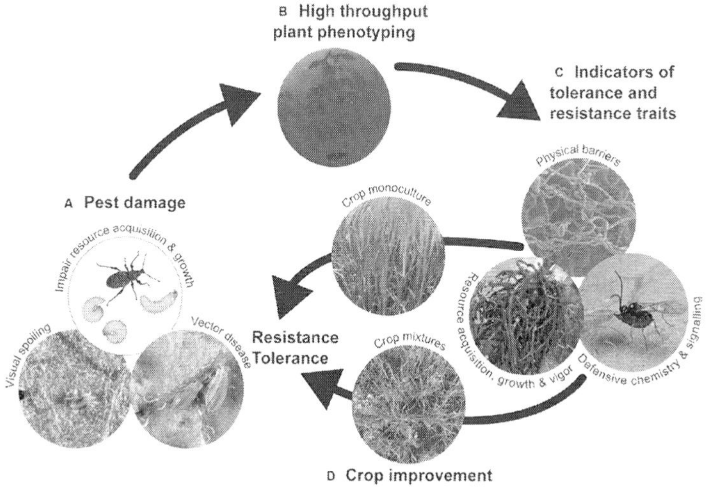

Figure 2.3. Plant Tolerance Mechanism against Pest Attack.

Source: Image by Pixabay

Tons of the research on the signaling occasions brought about via herbivores has focused on tomato and different solanaceous flowers. In this model system, the peptide device in acts at or close to the wound site to boost the manufacturing of Jasmonic acid.

Jasmonic acid or its metabolite's function phloem-cell long-distance alerts, and set off the expression of protection genes in distal elements of the plant (Belete, 2018).

2.2. DIRECT MECHANISM OF DEFENSE AGAINST PESTS

Plant structural tendencies including its specialized structures like thickness of cell wall, cell wall lignification content leaf wax, and trichomes form the primary bodily resistance barrier to feeding with the aid of the bugs, and the secondary metabolites such act as toxins and prohibits the pest growth and make further protective boundaries that guard the plant against next attacks.

Other than that synergistic impact among one of a kind; defensive additives improve the protection system of the plant life towards the invader.

For instance, when oxidative enzymes, phenolic compounds, Proteinase Inhibitors, alkaloid secondary metabolites in tomato are when ingested one by one resulted in limited impact; however, when acting together in a synergistic way they affect the insect at some stage in ingestion, digestion and metabolism. Beet armyworm infection in *Nicotiana attenuate* is inhibited when nicotine expression is synergistically combined with the trypsin proteinase inhibitor (War et al. (2012).

Insect herbivores belonging to any feeding guilds should make an interaction with the plant surface with a purpose to set up them at the host plant. It is far consequently now not unexpected that physical and chemical features of the plant's uppermost surfaces are critical causations of resistance. All plant parts offer a few sorts of resistance against herbivory. They range from tissue hardness to extraordinarily complex glandular trichomes.

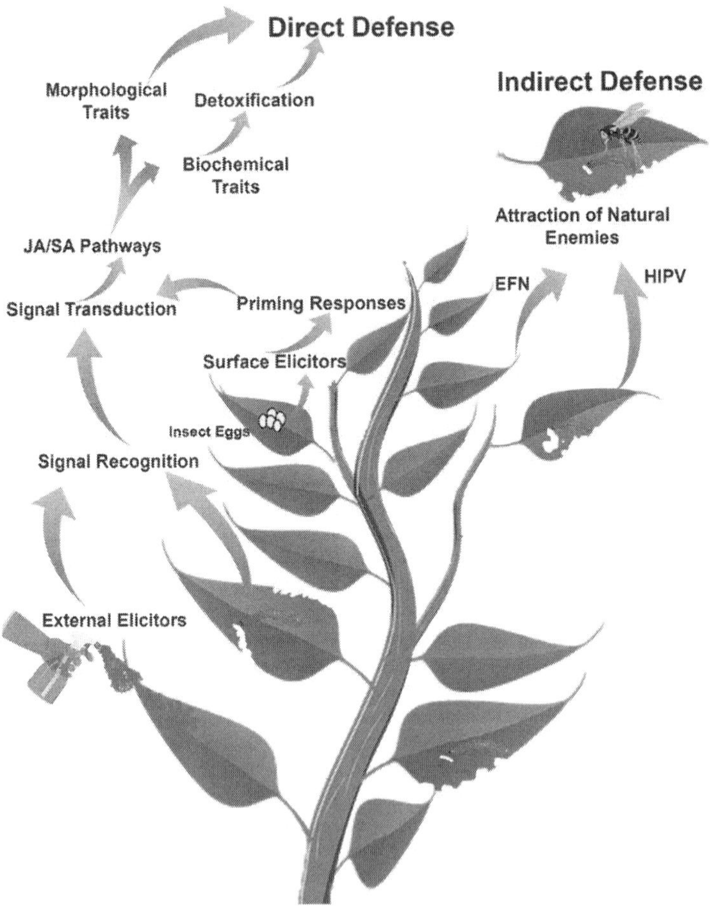

Figure 2.4. Plant Tolerance Mechanism against Pest Attack (Direct and Indirect Mechanisms).

Source: Image by pxhere.com

Cuticle along with wax perform their crucial function in desiccation tolerance, in addition, they enhance the slipperiness to great extent, which impedes the capacity of many non-specialized insects to populate leaf surfaces. The major content of the wax layer as well as its chemical composition are critical elements of primarily performed resistance. Primarily based on distinct findings, plant structures are the first line of protection against insect pests, and play an essential function in host plant resistance to bugs and insects.

The first line of plant protection in opposition to insect pests is the erection of a physical barrier either thru the formation of a waxy cuticle, and/or the development of spines, setae, and trichomes.

Structural defenses include morphological and anatomical tendencies that confer a health gain to the plant by way of at once deterring the herbivores/ insects from feeding and range from distinguished protuberances on a plant to microscopic modification in cell wall thickness because of the addition of lignin and subrin that leads to the lignification and suberization, respectively.

Structural traits including spines and thorns as in family spinescence and hardened leaves are known as sclerophylly, which play a leading role in plant protection against insect pests.

Sclerophylly refers to the specialized structure leads to the hardening of leaves, and performs an energetic position in plant defense towards herbivores by means of decreasing the palatability and digestibility of the tissues, thereby, reducing the herbivore damage.

Family spinescence includes plant structures consisting of specialized protective structures known as spines, thorns, and prickles.

It has been suggested to guard the vegetation in opposition to many bugs. Pubescence consists of the layer of specialized hairs extending from the dermis of the above floor plant elements along with stem, leaves, and even fruits, and arises in numerous forms including immediately, spiral, stellate, hooked, and glandular (Belete, 2018).

2.2.1. Trichomes

The plant dermis is regularly sheathed by outgrowths known as trichomes. They are in all mostly occurring categories of land. They arise typically from epidermal tissue and then increase and differentiate to make specialized hair-like entities.

Trichomes play an imperative function in plant protection towards many insect pests and involve both poisonous and deterrent results. Trichomes density negatively affects the ovipositional conduct, feeding and larval nutrition of insect pests.

Further, dense trichomes influence the insect routinely, and intervene with the motion of bugs and different arthropods on the plant surface thereby, decreasing their entry to the leaf dermis.

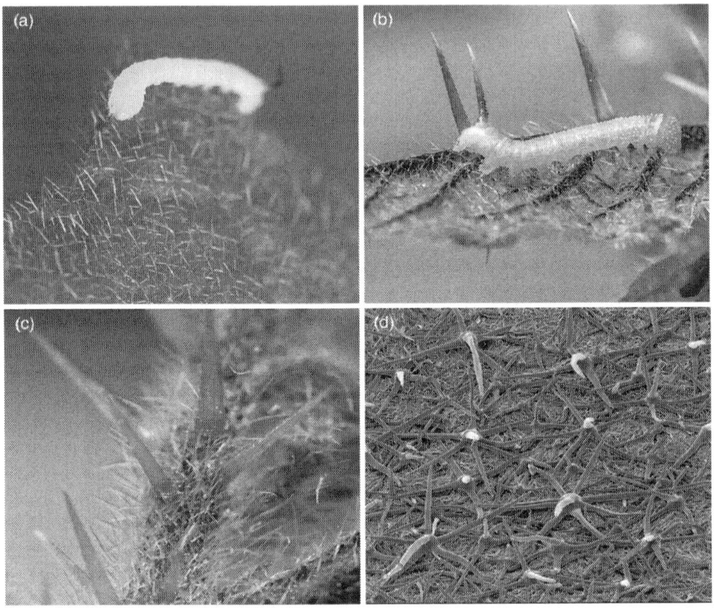

Figure2.5.Role of Trichome in Plant Defense.

Source: Image by onlinelibrary.org

The trichomes come in different shapes and can be glandular as well as non-glandular. They can be:

- Branched and unbranded
- Straight
- Spiral or
- Hooked

Glandular trichomes effectively synthesize and release secondary metabolites such as alkaloids, flavonoids along with terpenoids that can be extremely toxic, repellent, or trap bugs and different organisms, therefore forming a mixture of structural and chemical protection.

The induction of trichomes in response to insect harm has been reported in lots of plants(Agrawal et al., (1999). This growth in trichomes density in response to damage can most effectively be found in leaves developing throughout or next to insect attacks because the density of trichomes of existing leaves does not change.

Belete (2018) suggested that damage by means of adult *Phratora vulgatissima* in *Salix cinerea* plant caused augmented trichome density inside the new leaves growing thereafter. Likewise, an increase in trichome density in *S. cinera* in reaction to coleopteran destruction has also been discussed. Increase in trichome density after insect damage has also been mentioned in *Raphanus raphanistrum* L. In black mustard, trichomes density and glucosinolate degrees had been elevated after feeding by the tiny white butterfly.

2.2.2. Plant Phenolic Compounds

Amid the secondary mctabolites, plant phenols represent one of the most not unusual and large set of plant shielding compounds, which play a chief role in the resistance of the host plant towards natural pests.

Phenols are plant secondary metabolites concerned in plant defense against insect herbivory and are synthesized as monomeric and polymeric phenols and polyphenols from shikimate- phenylpropanoids - flavonoids pathways (Zhao et al. 2009). Phenols perform their function through a protecting mechanism not only towards insects, but however also in opposition to microorganisms and competing plant species like weeds.

Lignin is one of the most famous phenolic heteropolymers that plays a principal function in plant defense against hazardous pests and pathogens. It limits the access of pathogens with the aid of blocking off physically or growing the leaf sturdiness that reduces the feeding by using bugs, and decreases the dietary content of the leaf. Lignin synthesis has been found to be induced with the aid of insect or pathogen attack and its fast deposition reduces to increase of the pathogen or insect richness and prolificacy. Phenols are right now toxic to bugs and or act as feeding deterrents.

In addition, some phenols appeal to natural enemies of the insect pest. Phenolic compounds collectively with cucurbitaceous which is probably sour in flavor and make vegetation hostile to a good-sized variety of herbivores, which consist of mites and vertebrate grazers (Tallamy et al. 1997).

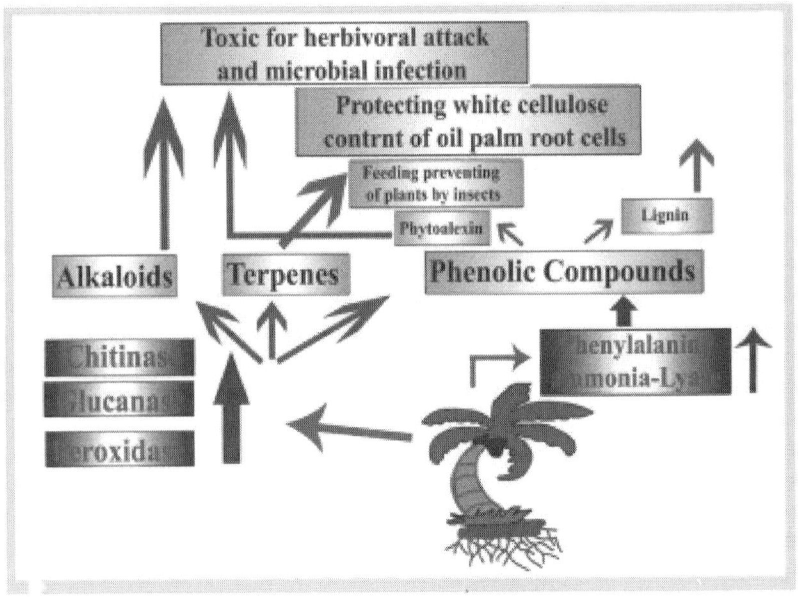

Figure 2.6. Role of Phenolic Compounds in Plant Defense.

Source: Image by Wikimedia

2.2.3 Role of Secondary metabolites In Direct Plant Defense

Secondary metabolites for chemical protection of plants vegetation produce massive and numerous arrays of natural compounds that appear to haven't any direct functions in growth and improvement i.e., they have no usually recognized roles inside the method of photosynthesis, respiratory, transport shipping, translocation, nutrient assimilation and differentiation.

These compounds or chemical substances play an enormous function in direct defense to impair herbivores with the aid of considered one of the fascinating mechanisms: those chemicals might also reduce the nutritional price of plant food, or they'll act as feeding deterrents or toxins. There was substantial debate as to which of those strategies is more essential for host plant choice and insect resistance.

A crucial part of this debate concerns the quantity to which version within the levels of number one and secondary metabolites have developed as a plant defense (Berenbaum et al, (1995). Plant metabolism, that's shared with hazardous pest and other dwelling organisms, constitutes carbohydrates, amino acids, and lipids as critical nutrients for the insect.

Secondary metabolites are the compounds that don't influence the ordinary growth and development of a plant, however, lessen the palatability of the plant tissues in which they may be produced.

The defensive (secondary) metabolites can be either constitutive stored as inactive systems or induced in reaction to the insect or microbe attack.

The previous are referred to as phytoanticipins and the latter as phytoalexins (antimicrobial compounds synthesized via vegetation that acquire swiftly at regions of pathogen infection). The phytoanticipins are particularly activated by using β-glucosidase all through herbivory, which in turn mediate the release of numerous biocidal aglycone metabolites.

The conventional examples of phytoanticipins are glucosinolates that can be hydrolyzed through myrosinases throughout tissue disruption that help to fight against the pest. Different phytoanticipins include Benzoxazinoids (BXs), which are broadly allotted amongst family Gramineae and provide an outstanding level of defense against harmful pests.

Hydrolyzation of BX - glucosides with the aid of plastid - targeted β - glucosidases throughout tissue harm leads to the manufacturing of biocidal aglycone BXs, which play an important function in plant protection against insects.

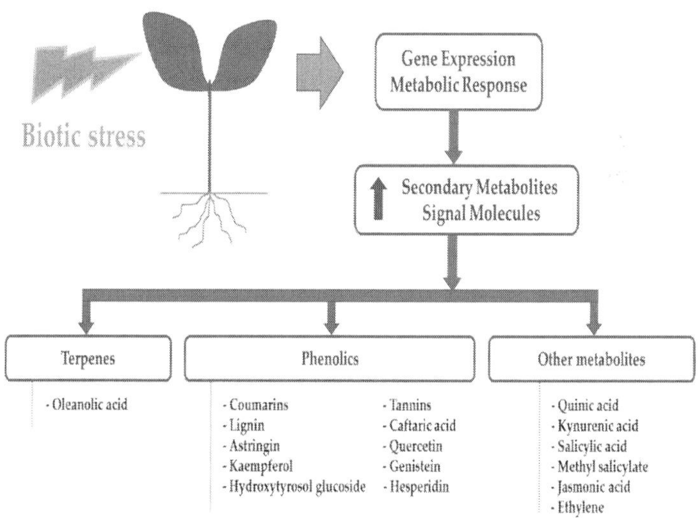

Figure 2.7. SecondaryMetabolites in Plant Defense.

Source: Image by flickr.com

Phytoalexins include alkaloids, isoflavonoids, terpenoids, which affect the overall performance and survival of destructive pests. The secondary metabolites now not useful to guard the flora from one-of-a-kind stresses, however, boost the fitness of the vegetation. It's been suggested that maize to corn earworm, *Helicoverpa zea* resistance is especially due to the presence of the secondary metabolites C-glycosyl flavone maysin and the phenylpropanoid product, chlorogenic acid.

Compound, 4, 4- dimethyl cyclooctene has been observed to be responsible for shoot fly resistance in sorghum. Looking at secondary metabolites may want to lead to the identity of the latest signaling molecules concerned in plant resistance towards insect pests (Belete, 2018). In the end genes and enzymes concerned with the biosynthesis of those metabolites could be recognized. Some of the secondary metabolites in plant defense may be subsequent.

2.2.4. Plant Defense Protein and Enzymes

Ecologically, in insect-plant interaction, the interrelationship is crucial for the survival of both. Pest and insects constantly search for a true and wholesome host plant that could provide them right meals and might be appropriate for mating, egg-laying and their deposition and provides feeding material for the offspring.

The nutritional requirements of harmful insects and pests are much like other animals and any imbalance in digestion and utilization of plant proteins by using the bugs' outcomes in drastic outcomes on the insect body structure.

Alteration of gene expression underneath stress including insect attack leads to qualitative and quantitative adjustments in proteins, which in flip play an important role in signal transduction, and oxidative defense.

Many plant proteins ingested by means of insects are strong, and continue to be intact within the mid gut, and additionally flow throughout the intestine wall into the hemolymph. An alteration in the protein's amino acid content or series influences the feature of that protein.

Likewise, anti-insect activity of a proteolysis- induced poisonous protein can be progressed by way of management of protease inhibitors, which prevent degradation of the poisonous proteins, and lets them exert their protective characteristics. Better information of protein structure and post-translational changes contributing to stability inside the insect gut would assist in predicting toxicity and mechanism of plant resistance proteins.

Latest advances in microarray and proteomic tactics have revealed that an extensive spectrum of plant resistance proteins is involved in plant defense in opposition to insects. Enzymes also one of the important factors of host plant resistance towards bugs is the disruption of insect's nutrition.

The enzymes that impair the nutrient uptake through insects via the formation of electrophiles consist of peroxidases, ascorbate peroxidases, and different peroxidases through oxidizing mono or di hydroxyphenols.

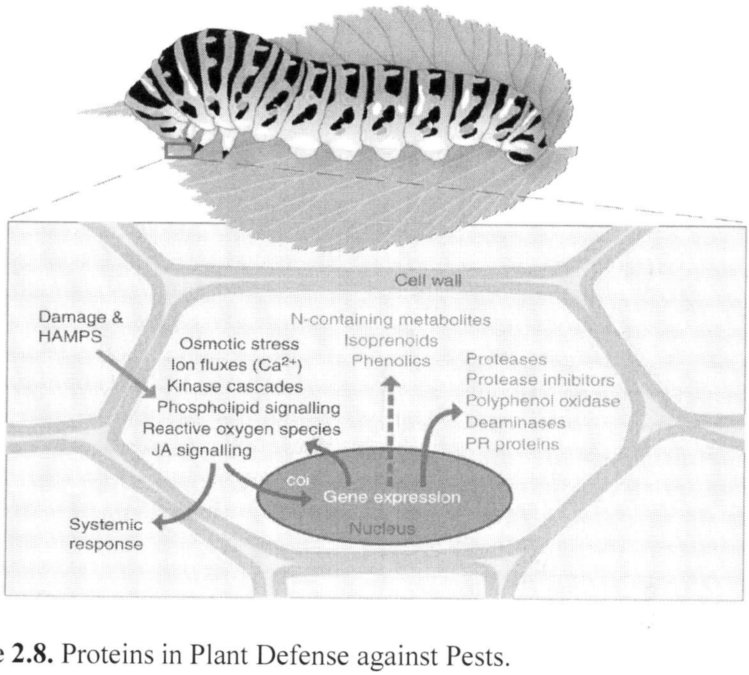

Figure 2.8. Proteins in Plant Defense against Pests.

Source: Image by flickr.com

Lectins are one of the glycoproteins, have protecting features in opposition to several pests. The insecticidal sports of different plant lectins had been applied as naturally secreted insecticides against insect pests. One of the most crucial locations of lectins is their survival within the digestive system of insects that gives them a sturdy insecticidal capability.

They act as antinutritive and/or poisonous substances by way of binding to membrane glycosyl groups lining the digestive tract, main to an array of dangerous systemic reactions. Lectins are solid over a massive range of pH.

Enzymes also one of the essential components of host plant interaction that act as a resistance against pests in order to disturb the insect's nutrition uptake processes. The enzymes that play important roles in plant pest defense are as follows:

- Peroxidases,
- Polyphenol oxidases,
- Peroxidises having the ability to oxidize mono or dihydroxyphenols,
- Ascorbate peroxidases

2.2.5 Proteinase Inhibitors

Proteinase inhibitors (PIs) cover one of the most abundant protecting functions of proteins in flora. Higher concentration of PIs occurs in storage organs together with seeds and tubers, and 1 to 10% of their general proteins include PIs, which inhibit special styles of enzymes and play a crucial position in plant defense against insects. PIs bind to the digestive enzymes within the insect intestine and inhibit their activity, thereby reduce protein digestion, resulting in the scarcity of amino acids, and slow improvement and/or hunger of the insects.

The protecting characteristic of many PIs in opposition to insect pests, at once or by way of expression in transgenic flowers to improve plant resistance in opposition to bugs has been studied in opposition to many lepidopteran and hemipteran bugs. The achievement of transgenic crops in expressing PIs against insect pests has accentuated the need to apprehend the mechanisms, and interactions of a couple of PIs with different defenses, and the adaptive responses of the bugs.

2.3. INDIRECT MECHANISM OF PLANTS AGAINST PESTS

The protective reaction in plants to attract their natural enemies of insects plays a pivotal position in shielding the plants in opposition to insect attack. Indirect defenses may be constitutive or brought about due to mixed motion of mechanical harm and elicitors from the attacking herbivore.

Production of volatiles and the secretion of more floral nectar mediate interactions of flowers with natural enemies of the insect pests through many kinds of interactions such as parasitoids or predators, which actively reduce the numbers of feeding herbivores. Induced indirect defenses have acquired growing interest these days and had been studied at the chemical, biological, genetic, and ecological ranges.

Herbivoreinduced plant volatiles play a vital role in plant defense by either attracting the natural enemies of the herbivores or by means of appearing as feeding and/or oviposition deterrent. Herbivoreinduced plant include lipophilic compounds with higher vapor pressure, which might be released from the leaves, flowers, and end into the atmosphere, and into the soil from the roots via plant life as a reaction of herbivore attack.

The Herbivore-induced plant produced range is in line with the plant and herbivore species, the developmental degree and situation of the flora and the herbivores.

A top-quality number of volatile compounds is usually released by way of the plants into the ecosystem, while a distinctive mixture of volatiles is produced in reaction to herbivory. The volatiles' combination launched through vegetation in response to insect attack is unique for a particular insect-plant gadget, which includes naturally occurring enemies and the neighboring plants.

The Herbivore-induced plants mediate the interactions between vegetation and arthropods, microorganisms, undamaged neighboring vegetation, or intraplant signaling that warns undamaged network inside the plant. Relying upon the modes of the feeding of insect pests, special defense signaling pathways are activated, which induce the manufacturing of precise volatile compounds.

The Herbivore-inducedplant includes terpenes, green leafy volatiles, ethylene, methyl salicylate and other VOCs. The properly-studied metabolites of hydroperoxide lyase generating pressure inducible compounds are the green leafy volatiles. Green leafy volatiles are reactive electrophile species types and defense signals. Green leafy volatiles consist of C6-aldehydes and their respective derivatives

To apprehend the role of C6-aldehydes and their respective derivatives (Maffei et al., 2010) in plant protection, the Green leafy volatiles degrees have been altered both via application of elicitors,96 or with the aid of manipulating genetically the HPL expression in plant life. They also play a crucial position in plant defense by using attracting herbal enemies.

Plant volatiles such as methyl salicylates and the C16- homoterpene has been determined to attract the predatory mites. The most frequent element of the HIPVs is methyl salicylate (MeSA) and has been mentioned in the headspace of many insect-infested floras inclusive of bean, and Arabidopsis. MeSA is a ubiquitous aspect of many leaf and floral blends and MeSA baited sticky cards attract many insect predators consisting of the big-eyed worm, ladybird beetle, and different natural enemies. One experimental study showed that inhibition of egg lying and deposition of cabbage moths *Mamestra brassicae* L. With the aid of MeSA launched all through infestation, suggesting that MeSA also can be detected by way of the attacking herbivores. Methyl benzoate (MeBA), which structurally resembles MeSA, has also been detected from insect-infested plants. *S. frugiperda* infestation in rice induces the emission of about 30 volatiles, consisting of MeSA and MeBA, which are particularly attractant to the natural enemies of *Cotesia marginiventris* (Pearse et al., 2020).

2.3.1 Definition of Indirect Plant Defense against Pest

Induction of defense as a plant trait that increases plant health by means of decreasing the bad consequences of herbivory or pathogen infection through adjustments within the nearby abundance or behavior of predators, parasitoids, parasites or pathogens is termed as an indirect plant defense against insects or pests.

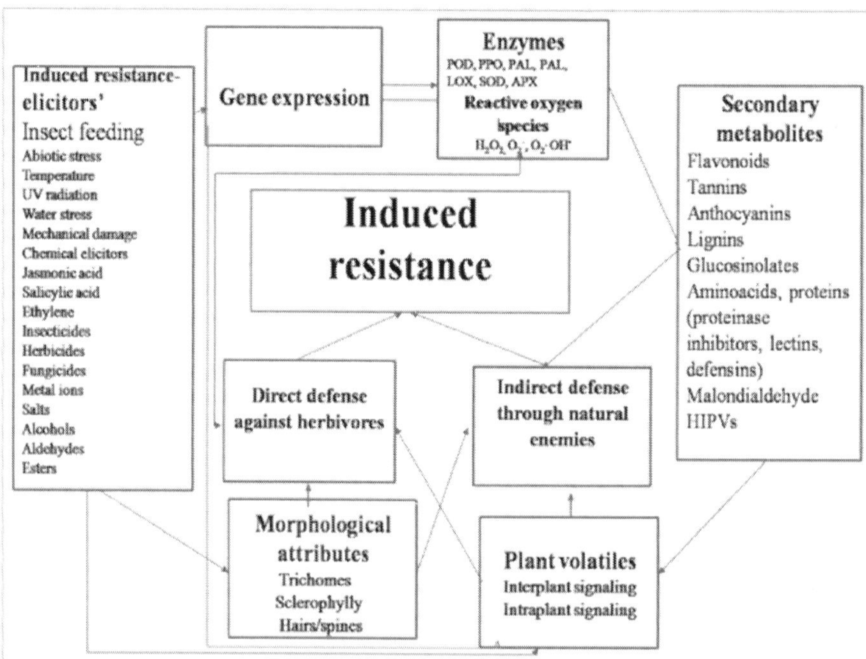

Figure 2.9. Mechanism of induced Resistance in Plants

Source: Image by Wikimedia

2.3.2 Herbivore Induced Plant Volatiles (HIPVs)

In this case, flowers do not directly guard themselves from insect feeding by emitting a mix of volatiles and non-volatile compounds. Insect-induced plant volatiles (HIPVs) play a vital function in plant protection via either attracting the natural enemies of the insects or by way of appearing as feeding and/or oviposition deterrent.

HIPVs are the lipophilic compounds with higher vapor pressure that are released from the leaves, plants, and result into the surroundings, and into the soil from the roots by way of flora in reaction insect attack. The HIPV's produced vary according to the plant and insect species, the developmental stage and conditions of the vegetation and the insects.

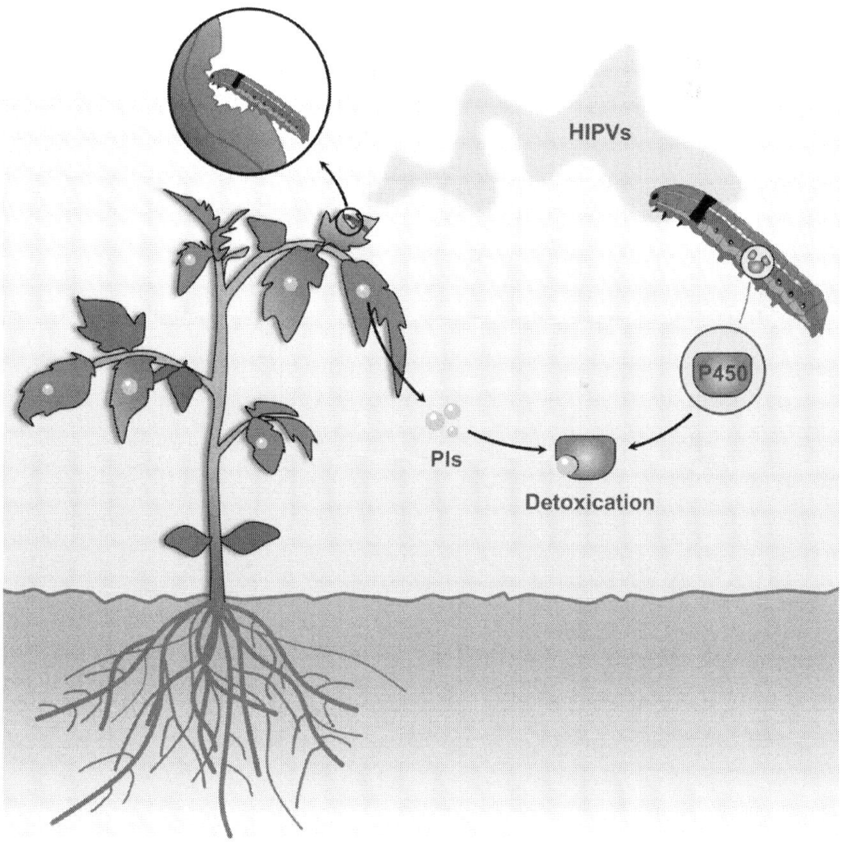

Figure 2.10. Herbivore induced plant volatiles (HIPVs).

Source: Image by Wikimedia

A premiere quantity of volatile compounds is commonly released by using the flora into the surroundings, while a one-of-a-kind combination of volatiles is produced in reaction to insect. The unstable combination launched through flora in response to insect attack is particular for a particular insect-plant device, which includes natural enemies and the neighboring flora.

The HIPVs mediate the interactions among vegetation and arthropods, microorganisms, undamaged neighboring plants, or intraplant communication that warns undamaged sites inside the plant. Depending upon the modes of feeding of insect pests, exclusive protection communication pathways are activated, which result in the manufacturing of precise volatile compounds (Walling et al., (2000).

Co-evolution of important and complicated plant-insect interactions performs a crucial role in structuring terrestrial ecosystems. Volatile chemical compounds emitted by using plants play key roles in communication among plants and different organisms.

Upon herbivore attack, flora growth emission of herbivore-induced plant volatiles (HIPVs) that defend flowers through eliciting each direct defenses towards herbivores and indirect defenses by using attracting herbal enemies which include parasites and predators. However, whether herbivorous insects, specifically insect larvae, can develop specialized counter-defenses in response to HIPVs is basically unexplored.

Here we reveal that caterpillars of the tobacco cutworm, one of the maximum vital insect pests of agricultural plants inside the Asian tropics, can perceive volatiles emitted from herbivore-challenged tomato plants.

Counter-defense responses elicited by using the volatiles encompass induction of eight genes encoding cuticle proteins, and seven genes encoding cytochrome P450s, and vital magnificence of cleansing enzymes.

Caterpillars therefore developed their performance on both herbivore-plant host plant life and when fed diets containing a toxin that interferes with the caterpillars' capability to digest proteins. There are three precise volatiles emitted by using host plant life named as:

- α-pinene
- Ocimene
- α-humulene

All these volatile compounds named above are associated with enhancing larval performance. Given that maximum terrestrial vegetation release HIPVs in reaction to insect herbivory, and insects are frequently uncovered to HIPVs, the capacity of herbivore insects to utilize those HIPVs to mount counter-defenses could play a pivotal position of their reaction to the induction of chemical-based defenses of host plant species.

2.3.3 Role of Elicitors

Plant life undergoes dynamic exchange throughout its course of life in response to herbivore brought about physical and chemical signs and indications such as insect oral secretions and compounds in the oviposition fluids. It is typically believed that insect-triggered plant responses are mediated with the aid of oral secretions and regurgitates of the herbivore.

The defenses generated by way of various elicitors differ primarily based on the type of the elicitor and the biological tactics concerned. An ability elicitor of herbivore-induced plant volatiles from the regurgitate of *Pieris brassicae* L. larvae has been recognized as β-glucosidase which ends up in emission of an unstable combo from automatically wounded cabbage leaves that appeal to the parasitic wasp, *Cotesia glomerata* (L.) that in turn control the larvae population to protect the plant against the pest (Belete, 2018).

At first, the term elicitor changed into used for molecules capable of inducing the manufacturing of phytoalexins, however, it's typically used for compounds stimulating any sort of plant protection (Hahn et al., 1996).

Finally, the induction of defense responses may cause greater resistance. In a broader meaning, the elicitors consist of both substances of pathogen origin termed as exogenous elicitors and compounds released from the infected plant by using the motion of the pathogen known as endogenous elicitors (Ebel et al., 1994).

Elicitors are divided into as following classes:

- Bodily or chemical,
- Biotic or abiotic
- Complex or defined

Current research has indicated tremendous similarities among the protection mechanisms prompted with the aid of fashionable elicitors and the innate immunity of animals, and it is tempting to take a position that the recognition of fashionable elicitors sooner or later ends in plant innate immunity.

Elicitors act as signal transduction compounds at low concentrations, imparting statistics for the plant to cause protection, distinguishing elicitors from pollution, which might also act handiest at better concentrations affect the plant detrimentally without lively plant metabolism. The advantages of elicitor treatments are as follow

- Reduced damage from insects, fungi, pests, and herbivores,
- Decreased environmental dangers as elicitors influence at once the crop plant, and their acute toxicity to different organisms is lower than that of insecticides, as shielding agrochemicals.
- Elicitors may be applied with the present-day spraying generation,
- Elicitor remedies could be an alternative to genetically modified plants for better appeal of herbal enemies of pest organisms on cultivated vegetation.
- Elicitor-treated flora undergoes decrease ecological dangers than GM flora (Poppy et all., 2005).

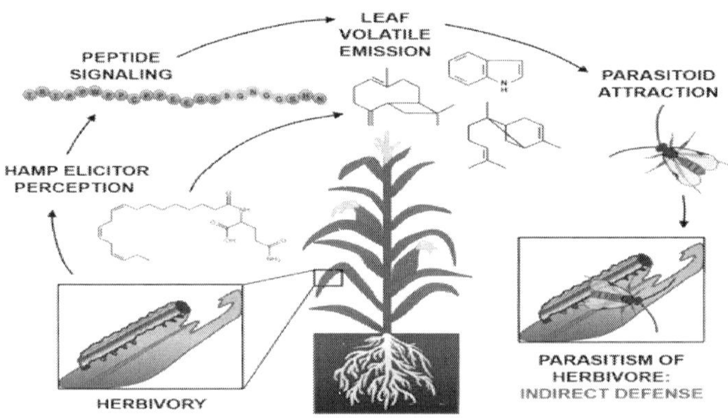

Figure 2.11. Role of Elicitors in Plant Defenses against pests.

Source: Image by biology.eused.org

2.3.4. Role of phytohormones in Induced Resistance in Plants

Plant protection against insect assault includes many signal transduction pathways which can be mediated by way of a network of phytohormones. Plant hormones play an important function in regulating plant boom, improvement, and defense mechanisms. Some plant hormones have been implicated in intra and inter plant linkages in plant life broken by means of insects.

Maximum of the plant protection responses against insects are activated through signal- transduction pathways mediated by using Jasmonic acid, salicylic acid, and ethylene. Precise sets of protection-associated genes are activated by these pathways upon wounding or by using insect feeding. Those hormones may also act synergistically or antagonistically, relying upon the attacker plant communication in vegetation broken by means of bugs.

Most of the plant defense responses towards insects are activated by signal- transduction pathways mediated through Jasmonic acid, salicylic acid, and ethylene. Units of protection-related genes are activated via those pathways upon wounding or by using insect feeding. These hormones may act in my view, synergistically or antagonistically, depending upon the attacker.

2.3.5. Salicylic Acid

Salicylic acid is a very important a remarkable phytohormone involved in plant defense against piercing and sucking types of insect pests. Responses to SA depend upon Non-expressor of Pathogenesis-related Genes1 (NPR1).

H2O2 prompted by way of SA in flowers defends them towards various insect pests considering H2O2 actively damages the digestive system of insects main to decreased growth and improvement SA signals, release of plant volatiles that entice the natural enemies of insect pests, e.g., Lima bean and tomato plants. MeSA serves as a volatile sign to triggerinduced defenses in flowers, such as HIPV emission, and some predaceous arthropods are interested in MeSA under field situations.

Salicylic acid (SA), a benzoic acid spinoff, is a critical phytohormone concerned with plant defense (Struggle et al., 2011). It is miles a critical endogenous plant growth regulator that generates an extensive variety of metabolic and physiological responses in vegetation involved in defense similarly to their effect on plant increase and development (Rivas et al., 2011). Responses to SA rely upon a regulatory protein referred to as Non Expressor of Pathogenesis-associated Genes1 (Pieterse et al., 2011).

The NPR1 gene is activated through redox pathways by means of SA accumulation and is translocated to the nucleus, however, it does not bind to DNA at once, however, acts through transcription elements. SA induces extra defense in opposition to piercing and sucking types of insect pests than the chewing ones.

SA signaling molecule is involved in neighborhood protection in addition to induction of systemic resistance. Manufacturing of ROS by using the SA pathway has been proposed to induce resistance in plants towards insect pests, e.g., in tomato vegetation against *H. armigera*. H2O2 induced with the aid of SA in flora defends them from diverse insect pests on the grounds that H2O2 actively damages the digestive gadget of bugs leading to decreased increase and improvement.

Furthermore, SA signals the release of plant volatiles that attract the natural enemies of insect pests, e.g., Lima bean and tomato flowers infested by means of spiders might attract the herbal enemies of spider mites. But it's been suggested that SA and JA act oppositely, where SA inhibits the activity of JA and vice versa.

MeSA serves as a volatile sign to trigger induced defenses in plants, inclusive of HIPV emission, and some predaceous arthropods are drawn to MeSA under field situations.

2.3.6. Ethylene

Ethylene is an extremely important phytohormone, which plays an active function in plant protection against many insects. Ethylene signal transduction pathway plays an essential role in prompted plant protection against herbivores and pathogens each at once and not directly, however, there are limited reviews on its position in indirect protection through the production and release of HIPVs (Yujie et al., 2016).

ET signaling pathway works either synergistically or antagonistically, with JA in of plant protection responses in opposition to pathogens and herbivorous insects. It is been suggested that ET and JA work together in order to extensity the effect in tomato for the PIs expression (Kunkel, 2002).

Infestation by way of *A. alni* brought about the emission of ethylene and release of numerous volatiles in Alnus glutinosa L. leaves similarly to mono-, sesqui and homoterpene.

ET precursor, 1-amino-cyclopropane-1-carboxylic acid has been mentioned to enhance the volatile secretion from the JA handled indifferent leaves. Ethylene similarlyinducedthe emission of volatiles caused by way of volicitin, JA in maize.

Even though diverse phytohormones are concerned with plant defense towards herbivores, JA is the most critical phytohormone connected to plant protection towards herbivores and activates the expression of both direct and indirect defenses (Usha et al., 2010).

2.3.7 Jasmonic Acid

JA is derived from linolenic acid via the octadecanoid pathway and accumulates upon wounding and herbivory in plant tissues.

Chewing of plant elements by bugs causes the dioxygenation of linoleic acid and linolenic acid through precise LOXs at C9 or C13 which might be transformed into 12-oxophytodienoic acid (12- OPDA) by means of allene oxide synthase and allene oxide cyclase. OPDA is transferred to the peroxisome, wherein its miles decreased by using OPDA reductase three (OPR3), forming JA. Oxidative burst produces ROS, which convert linolenic acid into phytoprostanes that sign trans- duction pathways.

A vast spectrum of protecting responses is triggered with the aid of jasmonate that consist of antioxidative enzymes, and trichome formation (Argandoña et al., 2001) A massive number of genes concerned in defense against herbivores are regulated by means of JA. The concentration of indole glucosinolate, and vital shielding compound, is caused via jasmonates. In addition to its role in the manufacturing of JA, OPDA signals the protection pathways personally. As an example, OPDA signaling regulates the COI1 -dependent and -independent transcription, changing the intracellular calcium degrees and cell redox potential. Jasmonates were found to engage with the COI1 unit of an E3 ubiquitin ligase complex which promotes binding of the COI1-unit to jasmonate ZIM-domain proteins, resulting in breakdown and degradation of JAZ proteins, which in any other case reduces and cease the JA inducible gene expression.JA has also been reported to affect calcium-based protein kinases (CDPK) transcript, and activity in potato flowers (Ulloa et al., 2010).

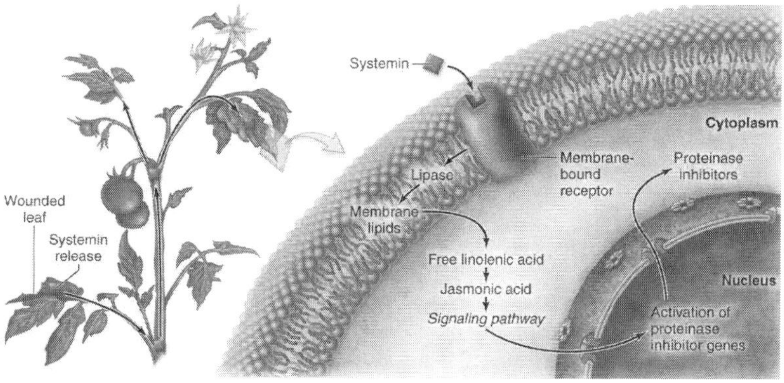

Figure 2.12. Role of JA in Plant Defenses against pests.

Source: Image by pxfuel.com

Table 2.1. Plant Specialized Compounds with insecticidal activity

Plant Insecticidal Compounds	Activity	Plant localization	Insect
C Based compounds			
Terpenoids			
Monoterpene alcohol	repellent	flowers	*Lasiusniger* (Hymenoptera)
Diterpenoids	repellentantifeeding	stems	*Ostrinia nubilalis* (Lepidoptera)
Cardenolides	toxicity	aerial and subterranean parts	*Danaus plexippus* (Lepidoptera)
Iridoid glycosides	toxicity	leavesnectar	*Junonia coenia* (Lepidoptera)
Phenolic glucosides	deterrenttoxicity	aerial parts	Generalist and specialist invertebrates
Aromatic esters	repellent	nectar	*Solenopsis xyloni* (Hymenoptera)
Flavonoids	repellent	leaves	*Spodoptera exigua* (Lepidoptera)
Isoflavones	feeding deterrent	roots	*Costelytra zealandicator* (Coleoptera)
Furanocoumarins and coumarins	toxicity	leaves	*Trichoplusia ni* (Lepidoptera)
Tannins	toxicity (oxidation)	leaves	*Orgyia leucostigma* (Lepidoptera)
N Based compounds			
Cyanogenic glucosides	toxicity	leaves	*Spodoptera frugiperda* (Lepidoptera)
Glucosinolates	toxicity	leaves	*Pieris brassicae* (Lepidoptera)
Alkaloïds	repellent	nectar	Bee pollinators
Pyrrolizidine alkaloids	toxicity	leaves	Non adapted Arctiidae (Lepidoptera)
Azoglucosides	toxicity (mutagen)	leaves, seeds, cones	*Rhopalotria sp.* (Coleoptera)
Non protein amino-acid	toxicity	leaves	Invertebrates

Protease inhibitors	toxicity	leaves	*Spodoptera littoralis* (Lepidoptera)
Peptides (cyclotides)	toxicity	leaves, flowers, stems, roots	Invertebrates

Source: Table by ndpi.com

2.4. GENE EXPRESSION

Large rearrangements in gene expression arise in flora in reaction to herbivory with masses, and even as much as several lots of genes getting up- or down regulated.

Advances in genomics and transcriptomics together with the availability of complete- genome collection statistics, expressed collection tags and complex microarrays, have led to better information of the modifications in gene-expression profiles in response to insect attack.

DNA microarrays provide a better and whole view of gene expression patterns and signaling responses mediated by insect elicitors and plant indicators, and have been demonstrated to be brilliant systems to screen the expression of thousands of genes concurrently.

A big number of genes are expressed through the aphid, *M. persicae* in comparison with caterpillar, *P. rapae* attack. Lepidopterans generally elicit adjustments of the expression of genes involved in glucosinolate metabolism in Brassicaceae, detoxing, cell survival, and signal transduction, at the same time as the aphids adjust the expression of genes involved in cell wall adjustments, oxidative stress, calcium-associated signaling, and glucosinolate synthesis.

2.5. TRANS GENERATIONAL INDUCED RESISTANCE TO PESTS

Biotic and abiotic stresses in plant life have been discovered to induce resistance now not only in the maternal flowers, but also inside the offspring's (Agrawal et al., 2001).

This maternally triggered resistance particularly termed as trans generational immunity has been observed to shield the progeny of plant life exposed to herbivory from insect pests, besides generating energetic seeds and seedlings (Agrawal et al., 2001).

However, there are only few reports on trans generational immunity of vegetation against insect pests. Wild radish plants, *R. raphanistrum* damaged via *P. rapae* or treated with JA produce offspring with high stages of triggered resistance to this insect. Arabidopsis plant life exposed to stresses such as, cold , heat and flood, resulted in a higher homologous recombination frequency and multiplied genome methylation, which in flip triggered the resistance to stress in the progeny (Boyko et al., 2010).

Maternal plants with low to intermediate degrees of herbivore harm may want to produce the seeds that might be extra full of life and seedlings which can be shielded against insect pests.

But, in addition, studies are required to recognize the genetic and molecular mechanisms of such signaling interactions.

Moreover, research on plant-insect interactions must be focused now not only on genetic effects, but also toward the epigenetic law of plant protection pathways and insect responses, because a huge body of evidence has been tested for cell siRNA indicators and inheritance of DNA methylation primarily based changes in gene expression.

There may be an awful want for in- depth studies on this challenge to make the most for pest management by manipulating the maternal ecology. Expertise of transgenerational changes caused resistance may be the solution to several difficult questions concerning the capability of vegetation to withstand herbivores (War et al., 2012).

2.6. CONCLUSION

Insect herbivores may be extensively categorized as having either chewing or pierce-and-suck feeding patterns. Each predicts the form of crop damage they target, with chewing insects inflicting probably huge mechanical harm, piercing causing and sucking insects causing loss of important nutrient elements and frequently act as a vector of many harmful diseases.

The chapter further concludes that the insects will have either an extensive host range the usage of many kinds of plants as a source of food term as generalists or they may have a greater restrained specialized range of hosts called as specialists. It also provides information about the fact that the greater the insect is specialized, the greater will be the dealing with mechanism of host defense.

REFERENCES

1. Belete, T. (2018) 'Defense Mechanisms of Plants to Insect Pests : From Morphological to Biochemical Approach'.
2. Gatehouse, J. A. (2002) 'Plant resistance towards insect herbivores: A dynamic interaction', *New Phytologist*, 156(2), pp. 145–169. doi: 10.1046/j.1469-8137.2002.00519.x.
3. Harun-Or-Rashid, M. and Chung, Y. R. (2017) 'Induction of systemic resistance against insect herbivores in plants by beneficial soil microbes', *Frontiers in Plant Science*, 8(October), pp. 1–11. doi: 10.3389/fpls.2017.01816.
4. Mitchell, C. *et al.* (2016) 'Plant Defense against Herbivorous Pests : Exploiting Resistance and Tolerance Traits for Sustainable Crop Protection', 7(July), pp. 1–8. doi: 10.3389/fpls.2016.01132.
5. Pearse, I. S. *et al.* (2020) 'Generalising indirect defence and resistance of plants', *Ecology Letters*, 23(7), pp. 1137–1152. doi: 10.1111/ele.13512.
6. War, A. R. *et al.* (2012) 'Mechanisms of Plant Defense against Insect Herbivores', 7(10), pp. 1306–1320.
7. War, A. R. *et al.* (2018) 'Review SPECIAL ISSUE : Using Non-Model Systems to Explore Plant-Pollinator and Plant-Herbivore Interactions Plant defence against herbivory and insect adaptations', (July), pp. 1–19. doi: 10.1093/aobpla/ply037.
8. Yactayo-Chang, J. P. *et al.* (2020) 'Plant defense chemicals against insect pests', *Agronomy*, 10(8). doi: 10.3390/agronomy10081156.
9. Walling LL (2000) the myriad plant responses to herbivores. J Plant Growth Regul 19(2): 195-216.130.135.138.
10. Kunkel et al. (2002) Brooks DM. Cross talk between signaling pathways in pathogen defense.; 5:325-31;
11. Pegadaraju *et al.* (2005) premature leaf senescence modulated by the Arabidopsis Phytoalexin Deficient4 gene is associated with defense against the phloem feeding Green Peach aphid. Plant Physiology 139(4): 1927-1935.
12. Agrawal et al., 1999 Induced responses to herbivory in wild radish: effects on several herbivores and plant fitness. Ecology 80(5): 1713-1723.

13. Agrawal et al., 1999. Transgenerational consequences of plant responses to herbivory: an adaptive maternal effect? Am Nat 2001; 157:555-69; PMID:18707262; http:// dx.doi.org/10.1086/319932.

14. Argandoña et al., (2001). Ethylene production and peroxidase activity in aphid-infested barley.; 27:53-68; PMID:11382067; http://dx.doi.org/10.1023/A:1005615932694.

15. Berenbaum*et al.* (1995) Turnabout is fair play: secondary roles for primary compounds. J Chem Ecol 21(7): 925-940.

16. Boyko *et al.* (2020) Transgenerational adaptation of Arabidopsis to stress requires DNA methylation and the function of Dicer-like proteins. PLoS One 2010; 5: e9514; http://dx.doi.org/10.1371/journal.pone.0009514; PMID:20209086.

17. Poppy *et al.* (2005) Gene Flow from GM Plants—A Manual for Assessing, Measuring and Managing the Risks, Blackwell Publishing, Oxford, UK, 2005.

18. Ebel *et al.* (1994) “Elicitors of plant defense responses,” International Review of Cytology, vol. 148, pp. 1–36.

19. Hahn *et al.* (1996), “Microbial elicitors and their receptors in plants,” Annual Review of Phytopathology, vol. 34, pp. 387–412.

20. Maffei *et al.* (2010) Site of synthesis, biochemistry and functional role of plant volatiles; 76:612- 31; http://dx.doi.org/10.1016/j.sajb.2010.03.003.

21. Pedras, *et al.* (2015) Plant chemical defenses: Are all constitutive antimicrobial metabolites phytoanticipins, 10, 209–218. [CrossRef]

22. Pieterse *et al.* (2004) the spider in the web of induced resistance signaling pathways; 7:456-64; PMID: 15231270; http://dx.doi.org/10.1016/j.pbi.2004.05.006.

23. Rivas *et al.* (2001) Salicylic acid beyond defense: its role in plant growth and development; 62:3321-38; PMID:21357767; http:// dx.doi.org/10.1093/jxb/err031

24. Tallamy *et al.* (1997) Cucurbitacins as feeding and oviposition deterrents to insects. *Environmental Entomology* **26**:678–683.

25. Ulloa *et al.* (2002) Jasmonic acid affects plant morphol- ogy and calcium-dependent protein kinase expression and activity in Solanum tuberosum. Physiol Plant 2002; 115:417-27; PMID:12081535; http:// dx.doi. org/10.1034/j.1399-3054.2002.1150312. x.

26. VanEtten *et al.* (1994) Two classes of plant antibiotics: Phytoalexins versus "phytoanticipins". Plant Cell 1994, 6, 1191–1192.

27. War et al. (2012) Mechanisms of plant defense against insect herbivores. Plant Signal Behav 7(10): 1306-1320.

28. Yujie *et al.* (2016) Role of ethylene signaling in the production of rice volatiles induced by the rice brown planthopper *Nilaparvata lugens*. Chin Sci Bull; 51:2457-65; http://dx.doi.org/10.1007/ s11434-006-2148-3.

29. Zaynab *et al.* (2018) *Role of secondary metabolites in plant defense against pathogens. Microbial Pathogenesis.* doi:10.1016/j.micpath.2018.08.034

30. Zhao *et al.*2009. Biochemical and molecular characterizations of *Sitobion avenae*-induced wheat defense responses. *Crop Protection* **28**:435–442.

Chapter 3

Signaling Mechanism during Plant Pest Interactions

CONTENTS

The chapter provides insights about the signaling pathway that is especially linked with defenses against herbivores viruses, pests getting their food through sucking, and biotrophic pathogens. The truth is all kinds of plant populations continuously and repeatedly meet destructive pests sequentially by enemies at some specific point or area. It also focuses on the fact that the plant natural enemies set off doubtlessly one path that adjusts the degree for ability in order to interconnect with plant protective and shielding signaling routes and pathways.

3.1. INTRODUCTION

Amid the millions of years of evolution, plant vegetation has advanced shielding strategies towards plant eating pests that have caused countless molecular interlinked mechanisms at each the genomic and metabolic ranges.

The potential of an organism to survive relies upon its potential to respond fast and efficiently to external stimuli and to broaden powerful and sustainable defenses. Those defenses are meant to without delay shield the plant with toxic compounds or by utilizing the mean of mechanical defense systems which include thorns and the functioning of glandular trichomes. It can also be done circuitously protect it via molecular communication that could attract predators or parasitoids of the plant eating insects or pests (Wu, 2010).

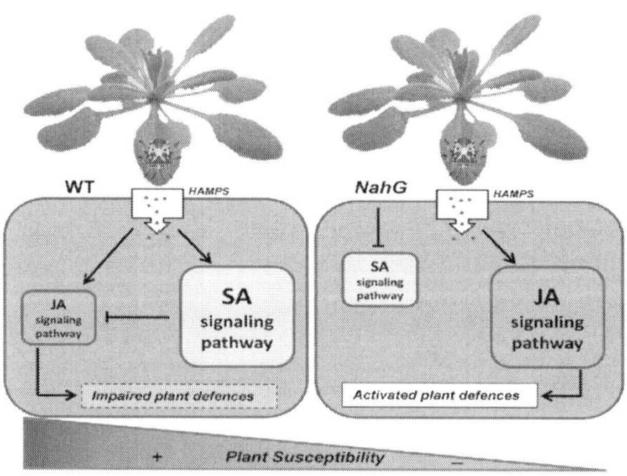

Figure 3.1. Signaling Mechanism during Plant Pest Interactions.

Source: Image by Luisa Ederli, 2020

The plant-mediated consequences are decided with the aid of plant brought about responses to harm through insects that eat plants or pathogens that undergone to changes and alternation of behavioral responses in a broad and extensive range of physical conditions as in the case of spines. The chemical defenses as in case of secretion of secondary metabolites, and the pathway of communicative signaling of plants associated through over representation of inducing and maintain defenses are goal to be enormously preserved among many plant species (Mithöfer & Boland, 2012) and that is the reason they probably lead to an obviously predicted reaction of plant initiated and triggered interactions amongst attackers. There are 2 most crucial pathways of signaling that are of hormonal nature interacting with triggered plant exhibition of shielding against destructive insects and pests are as follows: (Howe,2008).

- JA or Jasmonic acid
- SA or Salicylic acid

The activation of JA-based pathway of signaling is the goal to be achieved as defense in opposition to nectrophic pathogens, bacteria, chewing as well as mining plant eating pests, and nematodes.

Certainly, a prominent group of research has confirmed the sequential harm that is caused by pest attackers on the plant. It includes generating suitable and distinct hormonal pathways in a plant that may additionally bring about antipathy. It is also as prominent and distinctive as crosstalk between the special and fascinating signaling pathways, with this, in the end, determining the vastness and signal of results of initial pest on successive ones (Moreira, Abdala-Roberts and Castagneyrol, 2018).

As a result, the crosstalk speculation that predicts interference among signaling routes of the plant when they get attacked initially and later stages of attack bring about defenses related to extraordinary pathways of communication (Erb, 2012).

Particularly, the initial attackers, which induce the defences through the JA pathway, need to have negative effects on subsequent attackers, which induce the JA pathway but positive effects on subsequent attackers inducing the SA pathway. On the other hand, the initial attackers, which induce defences through the SA pathway, need to have negative effects on subsequent attackers, which induce the SA pathway but positive effects on subsequent attackers further inducing the JA pathway.

Some of research have determined styles that match the suggested antagonsim between JA and SA routes, both by immediately measuring plant growth regulating hormones or over representation of JA or SA associated defenses or chemical-based precursors, or by the way through herbivore or pathogen complete and detailed mode of action of bioassays of pathogens (Thaler et al., 2012). But, indications and signs of crosstalk are not always majorly wide spread and numerous researches have located effective interactions, this lack of consistency can be credited to several non-collectively specific parameters. The major and extensive studies showed that interactions among JA and SA-associated defense responses have used model plant include Arabidopsis or plants inclusive of tobacco or tomato which might not be a guide of action across a more species that are also non-domesticated ones, and other plant forms or organizations. Due to the fact plants respond with notably precise actions to specified living signs and indication, the nature and power in their reactions to stress may be firmly dependent on widespread factors of the living structures of plant enemies under study, including herbivores or even pathogens.

It may be related to purchaser developments which includes the food plan breadth within a given purposeful organization (Moreira, Abdala-Roberts and Castagneyrol, 2018).

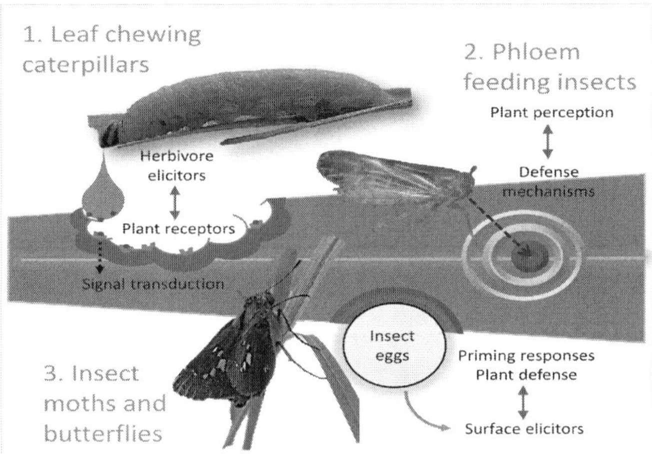

Figure 3.2. Defense Mechanism during Plant Pest Interactions.

Source: Image by Wikimedia

However, because of logistical obstacles, maximum research has not managed replicated experiments to govern the type of disease-causing pest attacker or developments of attacker groups.

3.2. DIRECT DEFENSE IN COMPARISON WITH INDIRECT PLANT DEFENSE

The plants have developed a broad range of defenses in order to compete with the attack of plant eating pests and insects. Historically, plant traits that at once affect herbivores had been investigated and explained. The supposed directly induced defense may also involve physical defenses through a modification into a thick cuticle, and chemical defenses like in the form of poisons and repellents. Defense generated and optimization chemical compounds care usually discovered in almost each principal instructions of plant based secondary metabolites as follow:

- Alkaloids
- Phenolics
- Flavonoids
- Terpenoids

Also, plants might additionally utilize defense generating proteins. Bodily and chemical-based parameters are frequently conjoined; for example, presences of toxic and sticky compounds as in the case of glandular trichomes and deterrents found within the upper layer of cuticle wax. In addition to the generation of defense mechanisms that are direct in nature, plants also have a capability to utilize other mechanisms to protect themselves.

Herbivores have to face a complex spread of natural enemies which includes predators as well as the parasitic wasps and vegetation can also amplify the effectiveness of those enemies and use them as saving agents (Dicke and Poecke, 2002).

Indirect defense protects and safeguards alternative feeding source chemical-based dataset, both by being alone or in communication with other (Wäckers, 1999). Occasionally, many plant lives have so-referred to as 'domatia' which might be systems used as safe heavens for carnivores, which include destructive insects like ants and pests for example predatory mites.

The natural enemies of pests that are now inhabitants of plants offer safety to the plant by casting off plant-eating pests. The plant also functions by making and offering floral or extra floral nectar that carnivorous arthropods get their food from and the synthesis of these nutrient materials may be prompted with the aid of herbivory (Koptur, 1992). Eventually, vegetation may also lure carnivorous arthropods with plant volatiles metabolized and synthesized in reaction to the plant eating insect's attack (Turlings, 1998). Leaves that remain undamaged of a plant release low levels of volatile chemicals, but while herbivorous insects damage a plant, the phenotype of the plant is absolutely modified and plenty of more volatiles are launched. This sort of phenotypic plasticity of vegetation in response to herbivory affects the community dynamics inside the ecological systems each without delay and indirectly for example through tritrophic interactions. The time period "phenotypic plasticity" is described as independent of antagonists or mutualists, when individuals of two species have interaction in a network, one partner can adjust its phenotype in response to its respective partner (Belete, 2018).

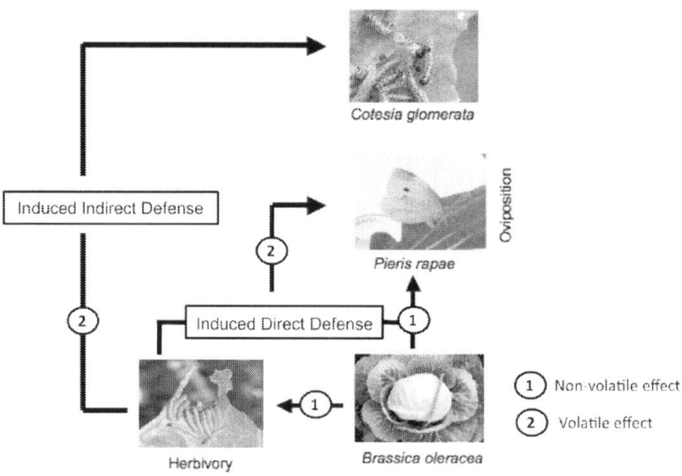

Figure 3.3. Representation of Direct and Indirect defenses.

Source: Image by bugsinthepicture.com

As a result, the overall appearance of the attacked plants emerges as tremendously appealing toward natural enemies that enhance their foraging closer to the damaged elements where the herbivores are feeding and that ultimately enhance the plant health with the aid of lowering herbivores.

Phenotypic plasticity in terms of "induced direct and indirect defenses" can affect many extra interactions inside the organisms belonging to extraordinary communities. The integrated understanding of herbivore induction, mechanism of volatile production and their principle at specific degrees of organisms is crucial for knowledge the ecological interactions and formula of manipulative systems in opposition to attacking herbivores.

3.3. SIGNAL TRANSDUCTION ANALYSIS

The signal transduction path and the route in question in the induction of direct defense had been properly studied and there is a fertile place for examining the cross conversation. Research on signal transduction in the induction of indirect defense is unexpectedly advancing and brushing space with knowledge in direct defenses. But, the two research fields have evolved independently, mainly with exceptional model systems. In direct defense, the extensive studied the plants of the Solanaceous family,

At the same time as in defense caused and induced induced, these include, for example, corn, and cabbage. It seems that there is a distinction between the solanaceous and cruciferous plants And this difference uses indirect defense (some of the novel compounds induced) in comparison to the plants such as Lima Beans and Maize (these are the major novel compounds which are induced by herbivory). To allow a greater evaluation of Transduction of signals in direct and indirect defense activated, the structures of the common model are noticeably necessary.

Numerous model systems seem to be correct candidates. Tobacco has been well studied for direct defense induced towards pathogens and insects and the transduction pathways in question (Baldwin, 1999). Lately, those studies have also incorporated indirect defense induction. This could make the tobacco a very interesting model of organism and system for comparative analysis.

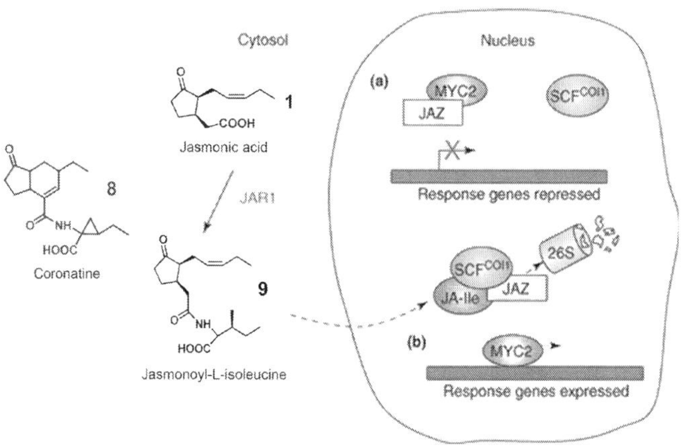

Figure 3.4. Representation of plant signaling pathway to induced JA.

Source: Image by flickr.com

There is an abundant understanding of the induction of direct defenses against pathogens and herbivores in tomato (Jones, 1994) and several mutants and transgenes are available. In addition, research in indirect defense induction is increasing. Consequently, tomatoes can be an exciting model device too. Arabidopsis thaliana has been studied properly for direct defense as opposed to pathogens. Research on direct defense against insects is increasing properly.

In the same way, there is abundant information on direct defense in different crucifers in opposition to herbivores and pathogens. Recently, the main demonstration of indirect defense has been made in Arabidopsis (Van, 2000). Along with the presence of various mutants and transgenes that could be affected in signs transduction pathways and close to genomic analysis, Arabidopsis becomes an important ideal vital model plant for comparative studies on signal transduction in direct and indirect defense (Dicke and Poecke, 2002).

3.3.1 Signal Transduction through JA and SA Pathway

Jasmonic acid (JA) is the promising by-product of the lipoxygenase pathway, additionally referred to as the octadecanoid pathway (Sembdner, 1999). The substrate linolenic acid initiates the lipoxygenase pathway and resulting products including JA, traumatin and a complex of unstable compounds (Z)- and (E)-three-hexenol that contains six carbons in its structure.

The pathway of JA biosynthesis might be integrally demonstrated, but can moreover be caused with the aid of e.g., injury, getting food through bugs, or utility of microbial cell wall additives.

Methyl jasmonate (MeJA) is a complex outcome of volatile Jasmonic acid, and may characteristic as an airborne communicative molecule of signal transduction (Karban, 2000).

MeJA can possibly be effortlessly converted to JA in the plant. Consequently, MeJA is covered in this phase on molecules of jasmonic acid. Jasmonic acid is an integral essential molecule in induced direct defense in opposition to insects in lots of plant species. The significance of jasmonic acid in lesion or injury caused defense responses have been examined through the fact that:

- The Exogenous importance and usefulness of JA or MeJA usually trigger and generate required defense responses.
- The rapid increase in the amount of endogenous JA synthesis

after injury and lesion production corresponds and match up with the elicitation of defense responses,

- The blockage in the JA synthesis machinery and pathway also hinders the induction of the defense responses (Baldwin, 1997).

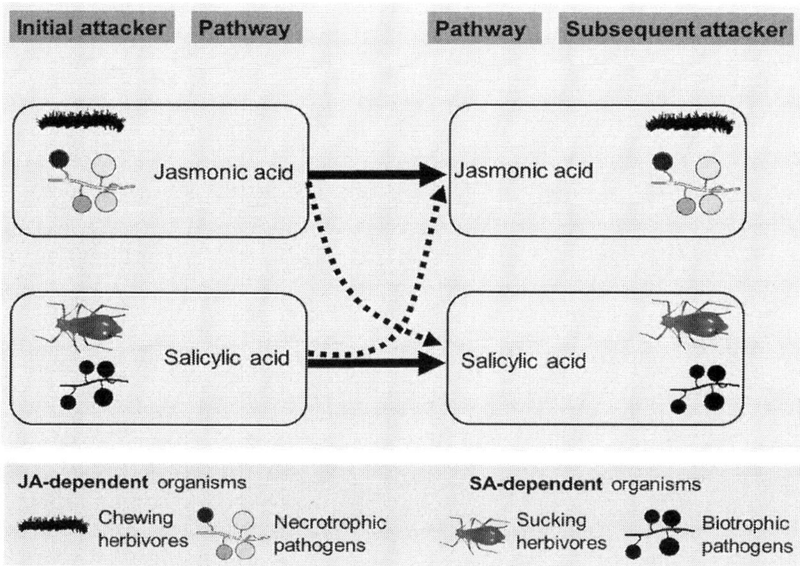

Figure 3.5. Representation of plant signaling pathway induced by an initial attacker.

Source: Image by (Moreira, Abdala-Roberts and Castagneyrol, 2018)

Moreover, transgenic plant and associated mutants have been critical systems to illuminate the importance and role of jasmonic acid in the transduction of cues and indications concerned in direct plant defense besides jasmonic acid, numerous different, related, oxylipins showed up to function as communicative signaling molecules as nicely those encompass:

Those include:

- 12-oxo phytodienoic acid and
- dinor-oxophytodieoic acid(Dicke and Poecke, 2002).

3.4. OXYLIPIN SIGNALING

It has been examined that in all plant species when a part of the plant gets injured due to pest attack resulted in the formation of the wound that causes a rapid deposition of JAs along with the launch of green leaf volatiles. An experiment in *Nicotiana sylvestris* and *Zea mays*, attack by way of adapted herbivores increase and enhance the formation of wound-induced JA accumulation that is initiated and trigger when secretions from the mouth of the feeding larvae are delivered into wounds throughout feeding.

Mouth secretions contain a complicated blend of potential elicitors, but solely a slew of fatty acid-amino acid conjugates have been proven to be vital and ample to elicit the JA enhancement and countless JA-dependent defense responses (Schmelz, 2003). Furthermore, eliciting leads to the signaling to allow JA synthesis, the mouth secretions from the herbivores are additionally recognized to increase the wound-induced GLVs secretions in *N. attenuate* (Pal, Scholar and Gardener, 2006).

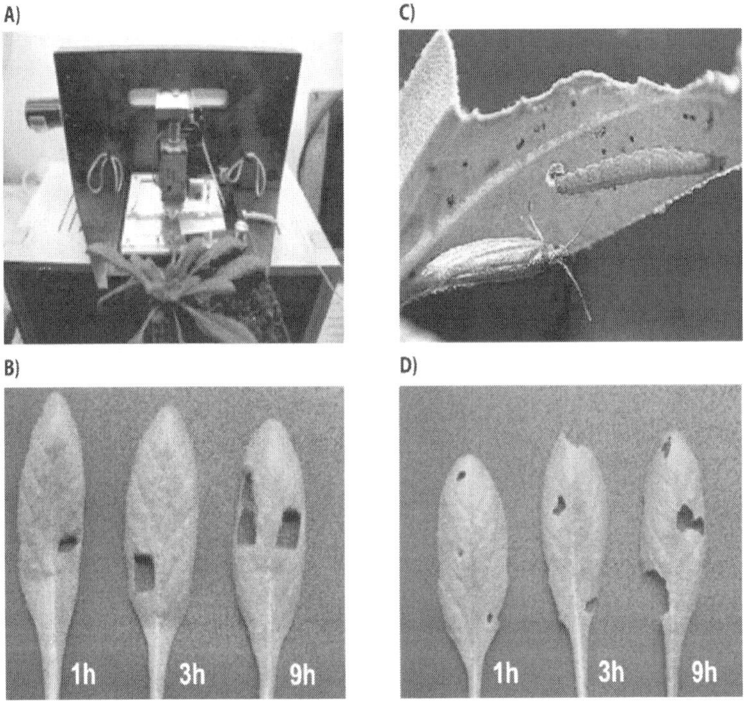

Figure 3.6. Comparison of the damage caused by MecWorm.

Source: Image by Dabrowska, 2014

The expansion and magnification of metabolite synthesis and gathering are no longer below transcriptional authority. Because lesions and wounds caused by plant attackers increase resistance as the release or accumulation of both JA and GLVs starts and goes side by side. It will increase and enhance the expression of oxylipin production genes. This mode of behavior among the gene expression and its associate product formation has been confirmed in *N. attenuata* via injury-causing lesions and deposition of allene oxide syntheses by herbivores.

Alongside hydroperoxidelyase (HPL) also is expressed which in turn is quite slower than the extra speedy and exceptionally transient expression in JA and GLV accumulation (Ziegler, 2001). The products that are formed are possibly maintained with the aid of the continuous changeable flux of fatty acid that is afterward metabolized through HPL that may be observed in healthy leaf tissues. The initial step in the production and synthesis in oxylipin formation, for instance, the manufacturing of JA and GLV is speeded up by the addition of catalyst name lipoxygenases that get incorporate a multi-gene unit with outstanding positional.

The path exhibits high specificity, leads to the development of tissue-specific gene expression and further localization of organelles to the high level of specificity. In potato plants, LOX-H1 accommodates the substrate for carrying out the HPL reaction process that is necessary for the release of green leafy volatiles, but it does not provide the hydroperoxide substrate for carrying out the octadecanoid route (Leon, 2002). One important example is of *N. attenuate* wounds that lead to the induction of more synthesis and accumulation of jasmonic acid that is dependent on the display of different molecules or homologs as in the case of NaLOX3.

The studies showed that DNA Silencing to inhibit expression of NaLOX3did now not affect or stop the green leafy volatile wound triggered synthesis and secretion. It also explained the interesting action of LOX forms that supply fatty acid hydro peroxide substrates to the Jasmonic acid and green leafy volatile transduction pathways. All these isoforms exhibit transit peptides of normal chloroplasts.

The 13-LOX products should be seen and present among the stroma and envelope fraction of chloroplast. The well-explained regulation of LOX further clarifies their role in providing the substrates for the specialized enzymes that are preferably focused to extraordinary chloroplast envelope outer membranes (Froehlich,2001). Surprisingly, the route specified supply of fatty acid hydro peroxides by distinct LOXs inhibits or stops the substrate crosstalk between the cascades of AOS and HPL as examined in *N. attenuata*.

The expression through antisense mediated silencing of HPL activity reduced the secretion and release of GLVs caused and triggered by wound and increased and amplifies the wound formation and herbivore-induced JA synthesis explosion respectively.

On the contrary, silencing of the above mention genes undertaking resulted in a reduction in accumulation in jasmonic acid and was go along with an increase in the release of HPL-based Green leafy volatiles (Halitschke, 2004). This crosstalk between substrates and corresponding enzymes could provide an outstanding regulatory mechanism to enhance or diminish individual elements of the complicated structured oxylipin signature during attacks from herbivores (Halitschke, Baldwin and Nteractions, 2005). Combination and unique blend of different concentrations of oxylipins could also influence unique and novel plant action and responses and this has resulted in the designation of oxylipin impression being crucial in the synthesis and regulation of plant defenses response.

3.5. ROLE OF SYSTEMIN IN PLANT-PEST INTERACTIONS

Systemin is an important 18-amino acid polypeptide and has been identified in tomato plants as a powerful persuader of protease inhibitors. Until now systemin have solely been observed in participants from the Solanaceae plant family, in particular bell pepper, tomato and potato (Constabel, 1998). Most researches for identification of systemin have been done on tomato. Lesions and injury trigger the systemic formation and deposition of systemin in tomatoes with the aid of an unregulated expression of systemin precursor called as pro-systemin specifically encoded by a specific gene (McGurl,1992).

As pro-systemin messenger RNA is systemically synthesized on the wounding, it is now not clear whether systemin is the kind of systemic induction of signal syntheses, inducing its own gene expression, or that every other systemic signal is being involved in it. It also has been demonstrated that systemin is carried away from the injured or lesion net site at some factor of the plant internally after 1 hour and 30 minutes since the wounding has been done. Other than that, systemin also induces prosystemin gene activity.

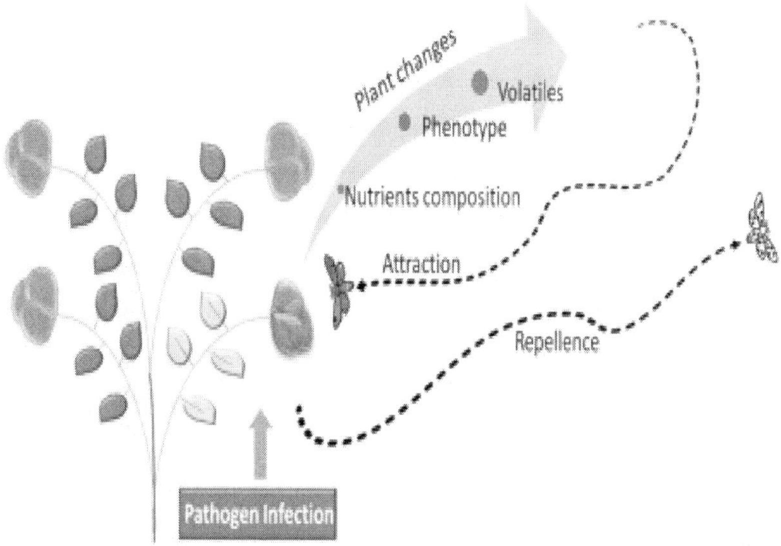

Figure 3.7. Plant-Pest Interaction and Signaled Synthesis of Defense Inducing Agents.

Source: Image by Flávia, 2017Dabrowska, 2014

Application of systemin through the reduce stems of tomato plant lifestyles precipitated the synthesis, gathering and deposition of PI proteins along with the very important mRNA. Transgenic tomatoes are now has been selected and made in which an antisense ProSystem in gene is induced and as a result a decreased systemin production is achieved that could be a good deal with much less protease inhibitor proteins are gathered after the onset of injury leading to the induction of wound.

These biotechnologically made transgenic plants of tomato had been extra inclined to feeding by *Manduca sexta* larvae as are more susceptible to the pest (Dicke and Poecke, 2002). Pin stands for Protein inhibitor genes, various tomato genes are induced by systemin. These embody genes encoding for other shielding proteins like:

- Polyphenol oxidase
- Lipoxygenase

Some proteolytic enzymes along with other important proteins.

These consequences show off that systemin performs a necessary role in the signal communication route of lesions, injury and wound triggered defenses mechanism in tomato.

3.6. ROLE OF CALCIUM IN SIGNAL MODULATION

In several communication and interactive pathways, changes and alternations within the cytosolic awareness of Ca2+ ([Ca2+] cyt) matches a large adjust concerning signals and responses. Calcium signal or the movement of downstream calcium-dependent activities plays a central role in the foundation of defense responses along with calcium-binding proteins.

Calcium ion is steadily employed a second messenger in numerous signaling pathways, shifting an extensive array of environmental or developmental stimuli in accordance with prominent physiological reciprocation. Under resting conditions, the cytosolic resting calcium level ([Ca2+]cyt) is maintained between 100 then 200nM which is 10^4 times less in comparison to that in the apoplastic fluid and 10^4 to 10^5 less in comparison to that in cellular organelles, further providing the potential for the ready import of Ca2+ into the cytosol where it acts as like a 2nd messenger.

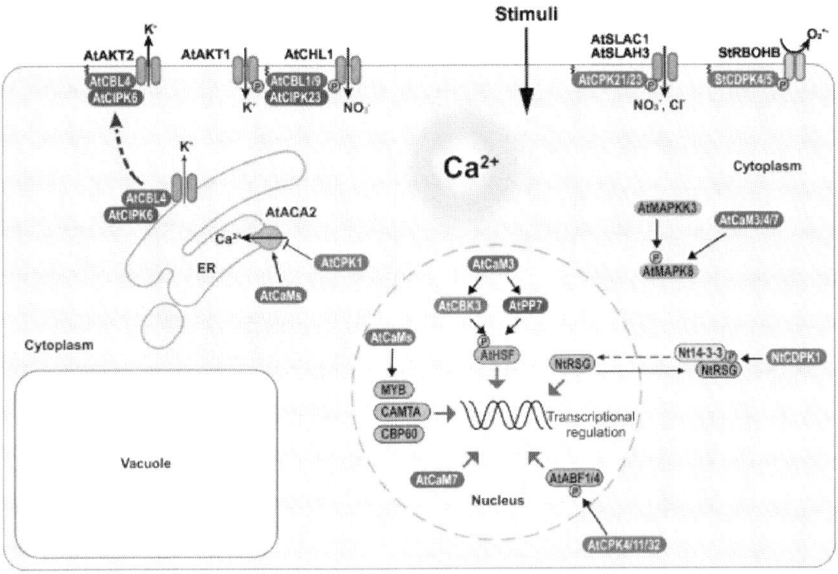

Figure 3.8. Role of Ca2+ in Plant Defense.

Source: Image by OliverBatistič, 2012

[Ca2+] cyt is carefully regular through the consistency about leaden fluxes (Ca2+channels) or lively conduct (Ca2+-ATPases along with the antiporters Ca 2+-) throughout the selectively permeable membrane and/ or endomembrane, then the buffering ability of the cytosol. Alternations into [Ca2+]cyt hold were pointed out for few indicators, inclusive of hormone growth regulators, light, abiotic forces and microorganism-based elicitors. How encoding along with decoding of the cell, the Ca2+signals evolved through a plethora of biotic yet abiotic stimuli is one about the just captivating questions into plant pest interaction biology.

During the past decades, studies over both animal as well as plant cells provide evidence that the spatial changes as well as temporal changes in [Ca2+]cyt (which is also referred to as 'calcium signature') that is further caused by a given signal tend to contribute to the specificity of the biological outcome (Hetherington, 2004).

These adjustments may additionally be on the increase so singular calcium transients, oscillations, or repeated spikes along unique sub-cellular mean point, deceleration time, spread and frequency.

Additionally, the same signal induces one of a kind Ca signatures rely on the organ, the tissue, or the cell type within speed (Moore ,2002) . [Ca2+] cyt Evolutionary domains remaininduced by stress regarding Ca2+ beside the extracellular medium, and by means of Calcium ion mobilization beyond cellular organelles, and/or with the aid of both.

The origin of calcium alerts might also be essential within the physiological action or response. The Information encoded into Ca2+ modification is through an order regarding Ca2+-binding protein spike thrust in conformity with a cascade regarding downstream effects, consisting of altered protein phosphorylation yet gene expression patterns.

Sensor relays such as calmodulin, and the calmodulin-related proteins yet calcineurin B-like proteins characteristic through bimolecular relationship. They endure a conformational alternate induced via Ca2+ before communicating with then changing the activity and structure of the proteins.

Sensor responders certain as like the Ca2+-dependent protein kinases (CDPK) function at first through intramolecular interactions or undergo a Ca2+-induced conformational change that alters the protein's very own activity or shape (Harper et al., 2004). These two modes concerning decoding calcium signals are used in a large amount among plant in order to provide a series of concerning regulatory modules allowing flexibility and diversity about biological responses triggered by calcium. Moreover, these calcium objectives must be kept close to calcium channels due to the fact of the low diffusion of calcium.

performed on tobacco cells with cryptogein treatment. Their facts suggest as NO promotes [Ca2+]cyt promotion via the assembly on intracellular pools on Ca2+, doubtlessly with the aid of cyclic guanilate monophosphate and cADPR- gated Ca2+-permeable channels.

It was investigated that the key position concerning NO manufacturing within grape cells treated with BcPG1. In this model, Nitric oxide formation relies upon [Ca2+]cyt, yet NO is concerned in [Ca2+]cyt homeostasis with the aid of regulating Ca2+ fluxes throughout the plasma membrane while activating Ca2+ release from internal stores.

Further research is wanted to look at Ca2+-permeable channels modulated directly yet in a roundabout way by nitric oxide production and understand the complexity of the interactions into Ca2+ and nitric oxide in signaling cascade which further result in plant defense activation (Lecourieux, Raneva and Pugin, 2006).

3.8. WOUND SIGNALING PATHWAY

The activities observed and discovered from wounding encompass ion fluxes across the selectively permeable plasma membrane, modifications in Ca2+ concentration present within the cytoplasm, are continuously alterinnng the ROS modifications in protein phosphorylation framework. These events lead to alterations in cell walls at the wound point, changing in consequence yet the generation of signals, which regulate protection gene expression.

The signals to that amount travel from broken cell membrane during the plant consist of pectic fragments derived beside the wall, contains Jasmonic, ABA and systemin. Even although mechanical wounding is brought about by means of an insect attack, are important differences into these twins' mechanisms about plant damage.

For example, the induction of wound leads to the amplification of JA levels is strengthened and promoted with the aid of herbivore feeding than by the application regarding oral secretions from larvae mouth and regurgitated to mechanical wounds, and allow the release of many well-known volatiles that attract parasitoids in an indirect protection mechanism (Fleurat-Lessard, 2015).

Different sheltering and protective mechanisms may responsible because of partial differences, however, different elicitors may additionally keep involved in order to safeguard the plant from pest harm. These resources are current within the oral secretion from the mouth of larvae of herbivores.

Two regarded pathways set off the synthesis or escape on volatile chemical alerts have been stated so far: a β-glucosidase by *Pieros brassicae* caterpillars N-(17-hydroxylinolenoyl)-L-Gln (volicitin) by army-worm caterpillars in beet. After documention of the harm, several not well-characterized events occur.

In essence, systemin is released within the vascular provision concerning damaged tissue activating and trigger the cascade of the octadecanoid signaling route. Many important reactions result in JA biosynthesis, on the regulation of the synthesis of signal pathway genes in the vascular bundles, then H2O2, the second messenger to activate putative defense genes and anti PIs genes in leaf mesophyll cells (Orozco-Cárdenas, 2001).

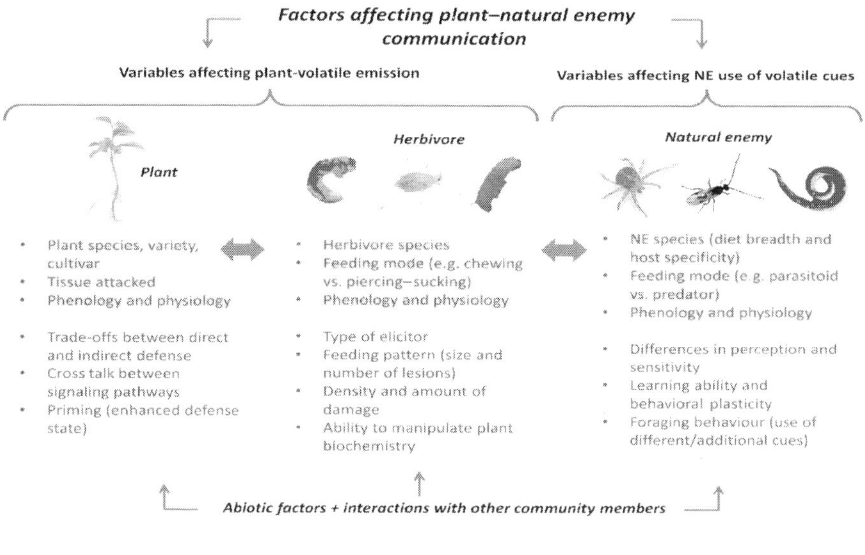

Figure 3.10. Communication between Plant-Natural Enemies.

Source:(Clavijo McCormick, 2016)

Despite the fact that JA is considered to be the predominant defense signal against chewing insects, the ethylene is considered to be an important defense modulator in a number of different plant species further acting concomitantly or even sequentially with the JA in receiver leaves.

3.9. GENE EXPRESSION IN SIGNALING PATHWAYS

The chemical aspects as well as the physical aspects of a plant are altered considerably which in turn follows phytophagy. The induced responses to

GMO based transgenic tomato exhibits the over production of the ProSystem in transgene bear been shown in conformity with assimilate a soluble cystatin constitutively. In leaves regarding wild-type tomato plants, cystatin mRNA amassed systemically in response in conformity with wounding, treatment along methyl jasmonate and systemin.

Moura, (2001) demonstrated up to expectation the induction concerning proteinase inhibitors in pepper in response to herbivore is regulated thru the octadecanoid transduction pathway. However, there are incomplete differences in the procedure among pepper now compared in simulation of the process previously described because of tomato plants (Pearce et al., 1991). During evolution, flora and bugs developed ecological, biochemical and physiological mechanisms parameters to deteriorate the impact over fly proteinases then plant PIs, respectively.

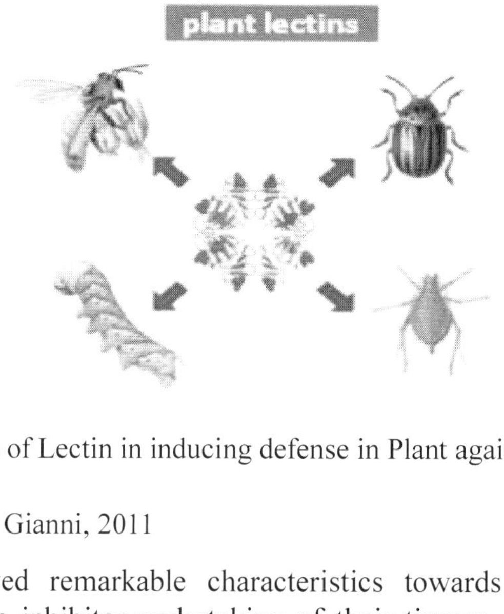

Figure 3.11. Role of Lectin in inducing defense in Plant against Pest.

Source: Image by Gianni, 2011

Plants evolved remarkable characteristics towards fly proteinases by enhancing the inhibitor undertaking of their tissues , with the aid of increasing a heterogeneous set of inhibitoractivities according to a range of enzymes involved (Christeller et al., 1998), by means of producing bidirectional inhibitors that are active on amylases and proteinases , by means of growing on the complex relationship of inhibitors together with exclusive biochemical pathways through the production over iso-inhibitors by forming exceedingly precise inhibitors on moth enzymes (Falco et al., 2001). In their turn, insects have found diverse ways of warding off negative effects induced by using proteinase inhibitors existing about their target plants. These incorporate the increase among the activity about their digestive enzymes and via the harmony about less sensitive enzymes (Paulillo et al., 2000), the amendment about the activities related to a variety of digestive hydrolases, the breakdown regarding the inhibitors in the gut through proteinases and thereduction of the sensitivity of their enzymes to the inhibitors. This is done by a rapid response which in turn up – regulates the gene expression of PI – insensitive proteinases so as to compensate for the inhibited ones (Brito et al., 2001).

Lectins are usually found in legume plants, in most cases within the storage organs and protective structures. *Phaseolus vulgaris* that is known as common bean includes a family of defense proteins such as:

- Phytohemagglutinins
- Arcelins
- α-amylase inhibitors

All of these are including the bean lectin gene family (Raikhel, 1991). The induction of insect toxicity by lectins has been examined and evaluated within unique insect orders. Vicilins, as associate according to the globulin family, are another classification of storage protective proteins located between leguminous seeds (Fleurat-Lessard, 2015).

3.11 CONCLUSION

The conclusion that can be drawn from the above chapter is that there are some specific interconnections within the various types of plants and multiple attackers that might further also impact the results of those changes. Also, these regularly vary amongst research studies.

It also concludes that the type of whether preliminary and successive pathogen and pest that feed at the identical or specific plant organ such as root and shoot, as plant triggered responses can be stressed out. The plant components vary their level of induction of resistance.

REFERENCES

1. Belete, T. (2018) 'Defense Mechanisms of Plants to Insect Pests : From Morphological to Biochemical Approach'.
2. Clavijo McCormick, A. (2016) 'Can plant–natural enemy communication withstand disruption by biotic and abiotic factors?', *Ecology and Evolution*, 6(23), pp. 8569–8582. doi: 10.1002/ece3.2567.
3. Dabrowska, P. A. (2014) 'OXYLIPIN PHYTOHORMONES IN PLANT-INSECT INTERACTIONS : ACTION AND METABOLISM OF JASMONIC ACID AND 12-OXOPHYTODIENOIC ACID IN PLANTS AND', (June).
4. Dicke, M. and Poecke, R. M. P. Van (2002) 'Signaling in plant-insect interactions : signal transduction in direct and indirect plant defence', *Plant Signal Transduction*, 3, pp. 289–316.
5. Fleurat-Lessard, F. (2015) 'Stored-Grain Pest Management', *Encyclopedia of Food Grains: Second Edition*, 4–4(figure 1), pp. 126–139. doi: 10.1016/B978-0-12-394437-5.00190-X.
6. Halitschke, R., Baldwin, I. T. and Nteractions, P. L. E. I. (2005) 'Jasmonates and Related Compounds in Plant-Insect Interactions', pp. 238–245. doi: 10.1007/s00344-004-0037-z.
7. Lecourieux, D., Raneva, R. and Pugin, A. (2006) 'Calcium in plant defence-signalling pathways', *New Phytologist*, 171(2), pp. 249–269.
8. Moreira, X., Abdala-Roberts, L. and Castagneyrol, B. (2018) 'Interactions between plant defence signalling pathways: Evidence from bioassays with insect herbivores and plant pathogens', *Journal of Ecology*, 106(6), pp. 2353–2364. doi: 10.1111/1365-2745.12987.
9. Pal, K. K., Scholar, V. and Gardener, B. M. (2006) 'Biological Control of Plant Pathogens', pp. 1–25. doi: 10.1094/PHI-A-2006-1117-02. Biological.
10. Baldwin, I.T. and Preston, C.A. (1999). The eco-physiological complexity of plant responses
11. Baldwin, I.T., Zhang, Z.P.,. and Schmelz, E.A. (1997). Quantification, correlations and manipulations of wound-induced changes in jasmonic acid and nicotine in Nicotiana sylvestris. Planta, 201, 397.
12. Brito LO, Lopes AR, Parra JRP, Terra WR, Silva-Filho, MC (2001) Adaptation of tobacco budworm Heliothis virescens to proteinase inhibitors may be mediated by the synthesis of new proteinases. Comp. Biochem. Physiol. 128B:365-375.

13. Constabel, C.P., Yip, L. and Ryan, C.A. (1998). Prosystemin from potato, black nightshade,
14. development of parasitic wasps.*Ecoscience*, **5**, 321.*Entomol. Exp. Appl.*, , (in press).
15. Falco MC, Marbach PAS, Pompermayer P, Lopes FCC, Silva-Filho MC (2001) Mechanisms of sugarcane response to herbivory. Genet. Mol. Biol. 24:113-122. Ferreira-da-Silva
16. Howe, A. G., & Jander, G. (2008).Plant immunity to insect herbivores. Annual Review of Plant Biology, 59, 41–66. https://doi.org/10.1146/annurev.arplant.59.032607.092825
17. Karban, R. (2011). The ecology and evolution of induced resistance against herbivores. Functional Ecology, 25, 339–347. https://doi.org/10.1111/j.1365-2435.2010.01789.x
18. Karban, R., Baldwin, I.T., Baxter, K.J., Laue, G. and Felton, G.W. (2000). Communication between plants: Induced resistance in wild tobacco plants following clipping of neighboring sagebrush. Oecologia, , in press.41
19. Koptur, S. (1992).Extrafloral nectary-mediated interactions between insects and plants. In
20. Mathieu Y, Kurkdijan A, Xia H, Guern J, Koller A, Spiro M, O'Neill M, Albersheim P, Darvill A. 1991. Membrane responses induced by oligogalacturonides in suspension-cultured tobacco cells. Plant Journal 1: 333–343.
21. McGurl, B., Pearce, G., Orozco-Cardenas, M. and Ryan, C.A. (1992). Structure, expression,
22. Mithöfer, A., & Boland, W. (2012). Plant defence against herbivores: Chemical aspects. Annual Review of Plant Biology, 63, 431–450. https://doi.org/10.1146/annurev-arplant-042110-103854Plant Mol. Biol., 36, 55.
23. Rakwal R, Agrawal GK, Jwa NS (2001) Characterization of a rice (Oryza sativa L.) Bowman-Birk proteinase inhibitor:tightly light regulated induction in response to cut, jasmonic acid, ethylene and protein phosphatase 2A inhibitors. Gene 263:189-198.
24. Reymond P, Weber H, Damond M, Farmer EE (2000) Differential gene expression in response to mechanical wounding and insect feeding in Arabidopsis. The Plant Cell 12:707-719.

25. Sembdner, G. and Parthier, B. (1993).The biochemistry and the physiological and molecular actions of jasmonates.Annu.Rev. Plant Physiol. Plant Mol. Biol., 44, 569.

26. Stotz, HU, Pittendrigh BR, Kroymann J, Weniger K, Fritsche J, Bauke A, Mitchell-Olds T (2000) Induced plant defense responses against chewing insects. Ethylene signaling reduces resistance of Arabidopsis against egyptian cotton worm but not diamondback moth. Plant Physiol. 124:1007-1017.

27. Thaler, J.S. (1999). Jasmonate-inducible plant defenses cause increased parasitism of herbivores. *Nature*, **399**, 686.to insect herbivores.*Planta*, **208**, 137.

28. Torres MA, Onouchi H, Hamada S, Machida C, Hammond-Kosack KE, Jones JDG.(1998). Six *Arabidopsis thaliana* homologues of the human respiratory burst oxidase (gp91phox). Plant Journal 14: 365–370. transposon tagging.*Science*, **266**, 789.

29. Wu JQ, Baldwin IT. 2009. Herbivory-induced signalling in plants: perception and action. *Plant, Cell & Environment* 32, 1161–1174.

30. Ziegler J, Keinanen M, Baldwin IT. 2001. Herbivore-induced allene oxide synthase transcripts and jasmonic acid in Nicotiana.

Chapter 4

Role of Phytochemicals and Phytohormones in Plant Defense against Pests

CONTENTS

It is now, well known that insects are originated during the Devonian period and since then co-evolution has been taken place with plants reaching a vast diversity in the context of their interaction with plants, so called insect-plant interaction. The chapter mainly focuses on the role played by phytochemicals and phytohormones in defending plants against pests and how it is useful in the safety of plants.

4.1. INTRODUCTION

Plants are continuously in battle with biotic and abiotic factors of their environment, and one of their major challenges is to protect themselves from the pests that eat them up, if not their population is not controlled at an initial stage. Plant pests and insects can lead to severe plant injury that ultimately affects its vigor and productiveness. Plants are continuously evolving to modify and enhance their protection and safety mechanisms in order to make them less prone to injury by pests and insects. It is mostly achieved through the production of certain chemicals by the plant called phytochemicals and plant growth regulating hormones.

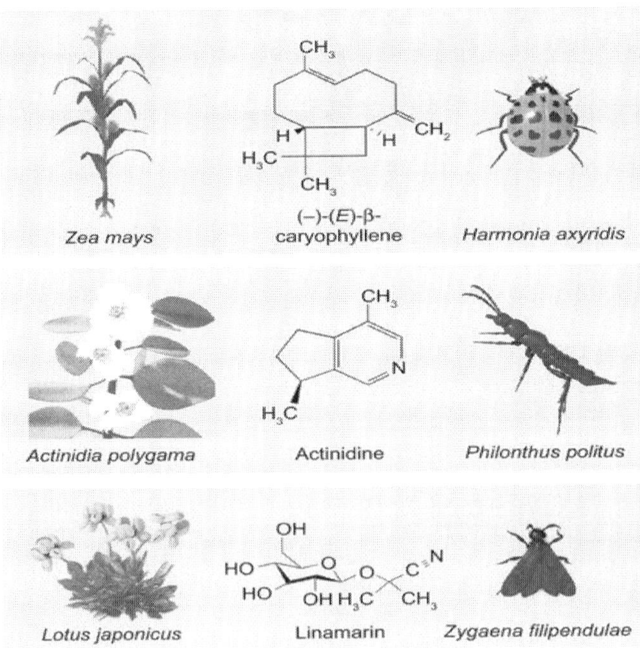

Figure 4.1. Chemical Convergence between Plants and Insects.

Source: Image by Franziska Beran, 2019

Chemical compounds are the by-products of secondary metabolism and are known as plant secondary metabolites. Same chemical compounds produce within a plant generates different kinds of protection mechanisms in different plants. In each variety of plant, phytohormones reduce the palatability, as in case of toxin, that acts as a stunne, or expand the expression of gene defense, amongst other plant species.

As plants are in constant war with their environment, growth regulating hormones and secondary metabolites act as a weapon to work against disease-causing agents. The plant chemicals and hormones also act as signaling messengers to generate the defense response against plant herbivores. Plants and pest interactions are the products of co-evolution that evolved over a thousand years in order to survive in challenging environments.

Both plant life and insects have evolved to have developed morphological and biochemical defensive traits to dodge their enemies' strategies. It is equally necessary to recognize the variations with the aid of insect pests to the plant protecting qualities in order to improve and set up managing techniques to overcome the insect pests.

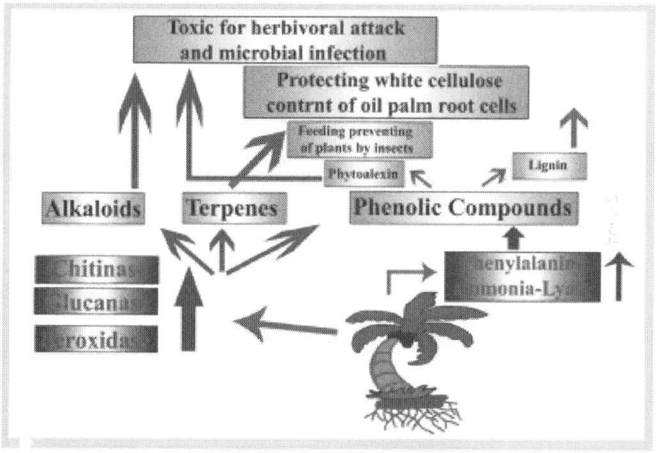

Figure 4.2. Production of Plant Secondary Metabolites on the Onset of Pest Attack.

Source: Image by Madiha Zaynáb, 2018

Co-evolution among plants and insects has been the point of attention and has been always considered as the criteria through which phytophagous insects come in the records of life for the very first time. Insects that are dependent on plants for their food needs form an imperative aspect in plant-pest co-evolution (Zhao et al. 2009). To avoid injury by using insect pests, plants have evolved to have an array of shielding strategies (Karban, 2011) through synthesizing more than a few physiological, morphological and biochemical programmed systems of defenses responses that avert the insect pests (War, 2012).

The morphological protective shielding responses include an increase in the amount of trichomes and enhanced latex deposition among the plant tissues (War et al. 2013). The development of biochemical defense features encompass several toxic secondary metabolites that are actively produced and released in plants as a response to insect attacks on healthy plants (Kaur,2015).

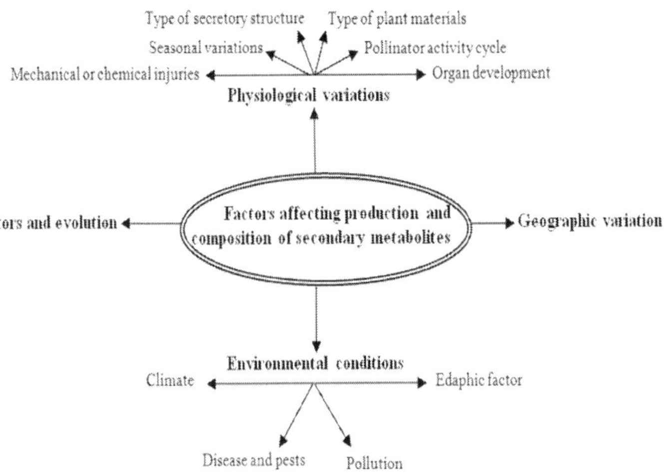

Figure 4.3. Factors Affecting the Production of Plant Secondary Metabolites.

Source: Image by Mansour Ghorbanpour, 2016

Plant defense might also directly influence insect growth and improvement through toxic secondary metabolites or indirectly with the aid of recruiting the naturally occurring enemies of the insect pest via induction of plant-based production and release of herbivore induced plant volatiles also known as HIPVs along with extra-floral nectar that attracts the natural enemies of plant pests (Karban, 2011). Induced resistance by the plant as a reaction to herbivore assault makes host plant vegetation phenotypically diverse and plant tissues much less nutritious, consequently making them virtually unappetizing to pests and insects (Karban, 2011).

Induced resistance is additionally experienced in the healthy parts of the infected plant and the neighboring vegetation has been also seen to have induced resistance against respective pest (Holopainen, 2013). Further, triggered resistance may additionally exhibit a trans - generational effect means it can be actively transferring to the offspring from their parents as reported in wild radish, *Raphanus raphanistrum* when it gets infected by *Pieris rapae* pest (Agrawal 2002).

The Trans-generational effect of brought about resistance makes flora greater vigorous and reduces insect infestation in upcoming generations. However, insects exactly adapt to the plant protecting features that enable them to feed efficaciously on the otherwise adversarial and uncertain host (Ahn et al. 2007; War, 2014).

This non-stop race between the two establishments to outsmart every other has led to the improvement of extra protecting characteristics in plant life and sturdy counter-adaptive strategies in insect pests (Sharma, 2009).

4.2. INDUCED PLANT DEFENSES AGAINST DESTRUCTIVE PESTS

Plants are designed to respond to herbivore attacks through a dynamic protection device that includes structural barriers, poisonous chemicals, and a series of steps to attract natural pest enemies. The defense process could be existing constitutively or brought about when any part of the plant gets injured and results in the formation of wounds by herbivores.

Most chemicals are produced as a reaction to herbivore attack. Induced defenses make the plants plastic physically, and a broad range of variability in defensive chemicals shapes up a better defense system. The induced defenses are usually introduced when previous or cutting-edge herbivory provides dependable indications of future attacks. The defense response is not easy to achieve and somehow is costly; whilst in environments is the place where herbivory is constantly high, and constitutive defenses need to be preferred. Herbivorous pests synthesize oral secretions containing chemically active ingredients that evoke plant reactions against herbivore insects and pests. The prevalence of plant elicitor production is widespread across wide-ranging plant families.

The plant produces diverse chemically distinct compounds, in response to herbivorous organisms. Some common examples are as follow:

- Fatty Acids
- Amino acids conjugates
- benzyl cyanide
- β-glucosidase
- Alkaloids
- Terpenoids (Huffaker,2007)

Plants have the tendency to identify herbivore elicitors and initiate a cascade of defensive reactions, inclusive of adjustments in plasma membrane capability and stimulation of complex meshwork of kinases and phytohormones.

Three main classes of plant growth regulating hormones that form a complex regulatory network essential in herbivore-induced protection responses are:

- Jasmonic Acid
- Salicylic Acid etc.
- Ethylene (Maffei,2007; Pieterse,2012)

Mechanisms of host plant resistance is an essential criterion for generation plant defense response in opposition to insect herbivory and is broadly implicated in crop protection approaches against insect pests and ailments (War et al. 2012). The naturally occurring constitutive plant protection mechanism is present in most plants regardless of the external stimuli, while the caused while the generated defense is encouraged via insect feeding or even by the elicitor application (Sharma et al, 2009). Additionally, Plant species can manage the sources between defense and growth by eliciting anti-pest defense only when quintessential (Karban, 2011).

A systematic review of genetic interdependence between plant resistances to multiple enemies has shown superb connection if each of the compared species are placed in the class of generalist and other is belong to the class of specific herbivores or plant eating insects (Leimu, 2006).

Mechanism of inducing resistive responses showed fine genetic interrelationship from herbivores with unique feeding habits, such as they may induce galls via gall inducers (War et al., 2018).

The process of resistance induction was examined in the pair grouped assessment between insect pests of special feeding guilds, such as dependance on phloem to get their feed and leaf-chewing herbivores, showed the lowest genetic connection. It is necessary to have both morphological and biochemical defenses in plants are required to withstand plant pest attack.

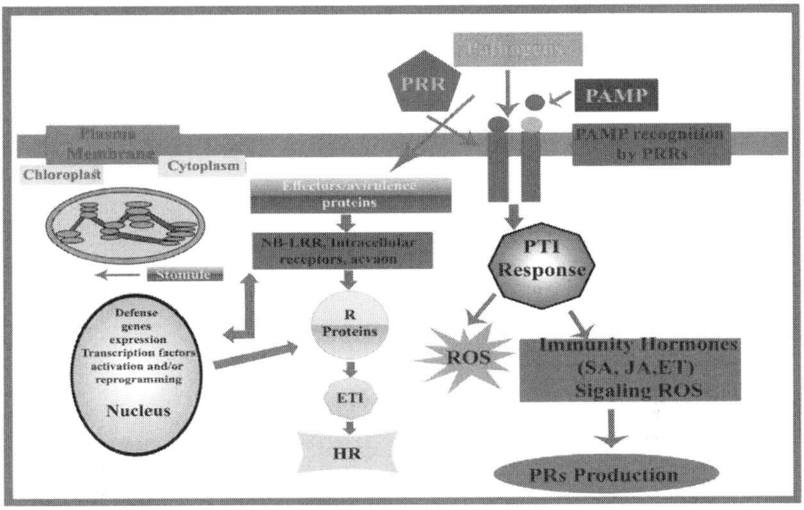

Figure 4.4. Defense mechanism of Plant Secondary Metabolites.

Source: (Safdar Abbas, 2018)

Although a morphological defense mechanism is particularly used by plant population against insect pests, the defense of biochemical origin is considered more effective as it directly affects insect population growth (Kariyat et al. 2013).

HIPVs are an outstanding way to protect vegetation with the aid of recruiting the natural or herbal enemies of the insect pests, for example as in the case of parasitoids and predators (War et al. 2011). Induced resistance in plant population against living stresses is contributed to the two pathways:

- Phenylpropanoid mediated by salicylic acid
- Octadecanoid pathways regulated and managed by jasmonic acid (He et al. 2011).

These pathways synthesized several plant protecting secondary metabolites that control and keep in check of pest population and actively secretes volatiles compounds that attract the insect's natural enemies to control pest growth and development (War et al., 2018).

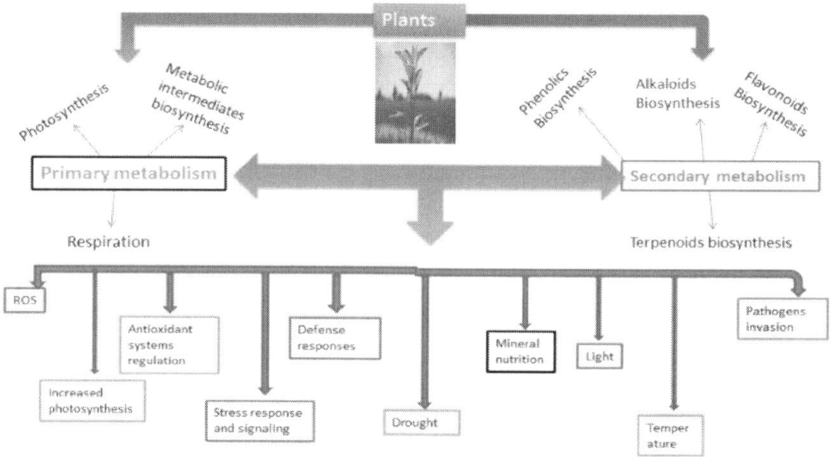

Figure 4.5. Stress and Plant Secondary Metabolites Production.

Source: Image by Tasiu Isah, 2019

4.3. PLANT BASED ELICITORS

Plants catch cues within the insect's oral secretion basically from their saliva and among the fluid of ovi-positional nature (Alborn et al. 2007). Mouth secretions of pests include unique elicitors such as fatty acid conjugates (FACs).

The forward drive into defense elicitor diagnosed beside the oral secretions in beet plant when beet armyworm, *Spodoptera exigua* attacks on it. It leads to the increase of a chemical known as volicitin that also is seen in maize plant at the onset of a wound. It resulted in the release of a combination of volatiles to that amount in order to attract the herbal enemies of the pest (Alborn et al. 1997).

N-linolenoyl-glu, is a plausible elicitor of volatiles into nicotiana plant. N-linolenoyl-glu is isolated from the tobacco hornworm, *Manduca sexta* regurgitate), so when applied after the wounded leaves of tobacco, it activates the mitogen-activated protein kinase (MAPK), along with some other important proteins such as:

- Wound-induced protein kinase (WIPK),
- SA-induced protein kinase (SIPK),
- Ethylene

- Jasmonic acid
- Salicylic acid
- JA-isoleucine conjugates (JA-Ile) (Wu et al. 2007).

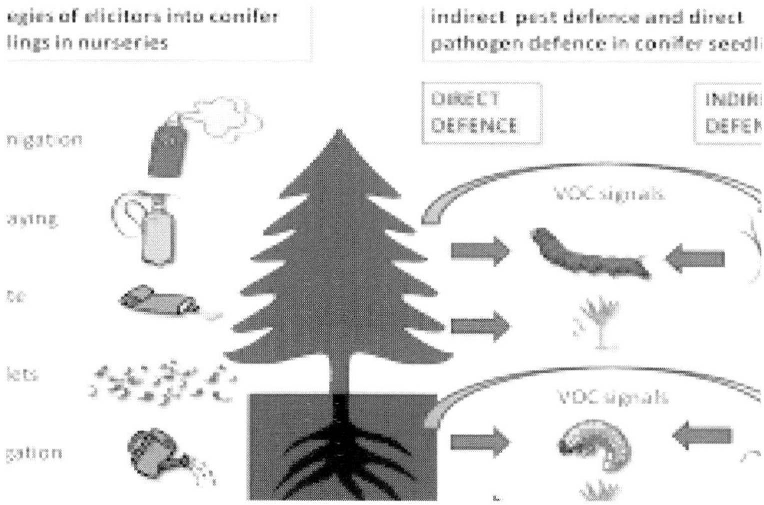

Figure 4.6. Plant Elicitors.

Source: Images of Holopainen et al., 2009

The MAPK access is also associated with an increase in plant population growth along with its development. It also enhances the quality of signaling transduction pathways among the host drive in compliance with biotic or abiotic factors as drought stress, cold stress and infections caused by pathogens other than insect attack (Wu, 2007).

Additionally, another 7-epi-JA induced by means of the usage of FACs elicits plant defensive genomic response toward herbivory. Glucose oxidase (GOX) in the case of *Ostrinia nubilalis* and *Helicoverpa zea* mediates the defensive signaling pathways in tomato (War et al., 2018).

Furthermore, oral secretion of *O. nubilalis* resulted in the expression of Proteinase Inhibitor 2 also known as PIN2 protease inhibitor genes in tomato and maize consecutively. Some reports exhibit the repression of plant protective responses by potentially regarding as insect-pest oral secretions. For example, oral secretions of African leaf worm of cotton, and cabbage butterfly, *Pieris brassicae* slow down the defense inducing responses in *Arabidopsis thaliana* ensuring the rapid increase in larval weights (Consales et al. 2012).

pest, the cell ruptures and releases secondary metabolite to induce resistance against pests.

Generally, glandular exudates comprise only one type of compound; however, this class varies amongst different species of plants. The glandular exudates commonly usually exist in the form of:

- Terpenes
- Resins
- Phenolics
- Alkaloids

4.4.1. Alkaloids

Alkaloids are grouped into several classes. This classification is primarily based on their heterocyclic ring system and biosynthetic precursor they made from. They include:

- Glycoalkaloids
- Imidazoles
- Indoles
- Isoquinoline purines
- Peperidines
- Pyrrolidines
- Pyrrolizidines
- Quinolizidines
- Tropanes
- Glycoalkaloids

These molecules are alkaline and incorporate nitrogen in a heterocyclic ring. Alkaloids are biosynthesized in plant roots majorly from amino acids and likely are involved in protection in opposition to insect herbivory. About 20% of vascular plants produce and secrete alkaloids, in particular in all plant families such as in Leguminosae, Solanaceae etc. (Sánchez-Sánchez, Morquecho-contreras and Morquecho-contreras, no date).

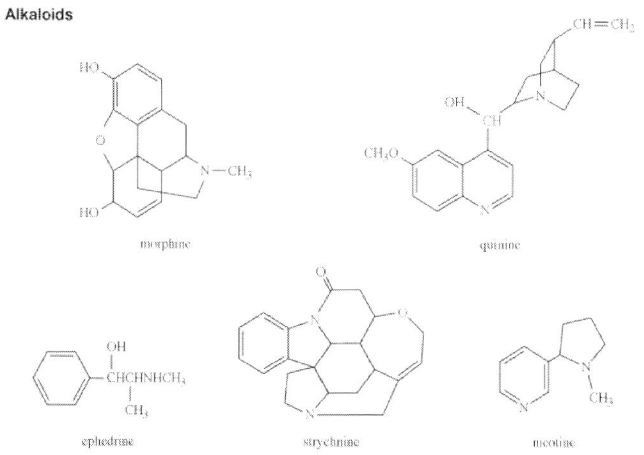

Figure 4.9. Chemical Structures of Alkaloids.

Source: (Holopainen et al., 2009) (Encyclopaedia Britannica, Inc)

4.4.2. Phenolic compounds

Phenolics are synthesized by plants to perform the following functions:

- Phenolic compounds tend to pushback herbivores,
- They can inhibit the enzymatic activity of plant pests to control their population
- Attracts pollinators as they have a very nice fragrance
- Help in fruit dispersers,
- Soak up UV radiation,
- Limit competitive behavior between plant neighbor's species (Lou, 2015).

Phenols are by-products of plant secondary metabolites concerned in plant defense mechanisms in opposition to pest attacks on a plant (War, 2012).

Phenolic compounds are synthesized as

- Monomerid
- polymeric phenols and
- polyphenols from shikimate -
- phenylpropanoids flavonoids pathways (Scott, 2010).

However, phenols are constitutively manufactured in plants, but their quantity depends upon the response to insect infestation as seen in coffee, wheat, *Ricinus communis* (Rani, 2011) and tomato.

Phenols are at once poisonous to insects and on the other hand may play their role as feeding deterrents (Dixit, 2017). Further, some phenols attract natural enemies of the insect pests to control their population. Phenolic compounds, for example like Cucurbitacins that are bitter in style and make plants adverse to a vast vary of herbivores, including worms, beetles, mites and vertebrate grazing animals (Balkema, 2003).

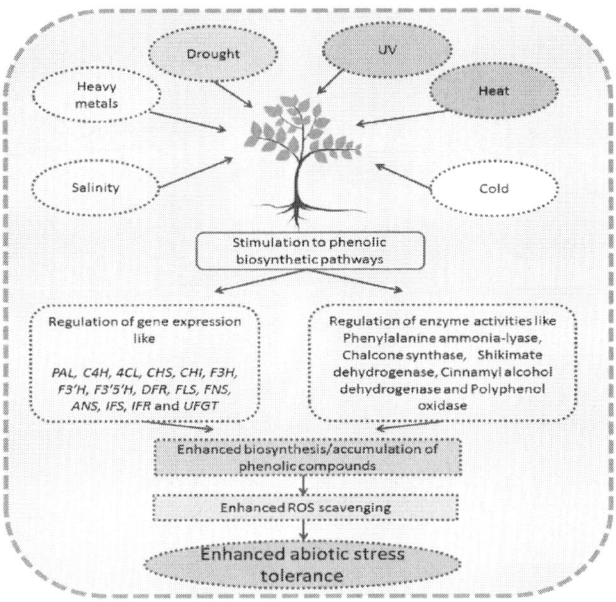

Figure 4.10. Role of Phenols in Plants.

Source: Image by Holopainen et al., 2009, Anket Sharma,2019

They either directly influence insect population and improvement (Balkema, 2003) or indirectly by appearing as ovi-position deterrents (Tallamy et al. 1997). Some reports advise that Cucurbitacins react as phagostimulants to insect pests. For example, *S. exigua* larvae expressed greater overall performance on *Cucumis sativus* genotypes with greater levels of Cucurbitacins than on the genotypes with decreased levels of Cucurbitacins (Barrett, 2004). Cucurbitacins B and D have been recognized as important phagostimulants for leaf beetles (Fukami, 1990). Also, insect pests that belong to the genus Aulacophora are recognized to sequester the cucurbitacins very well (Sánchez-sánchez, Morquecho-contreras and Morquecho-contreras, no date)

There are about 10,000 plant phenolics compounds isolated till now. Phenolics have the tendency to make covalent bonds with plant-eating pest digestive enzymes and inactivate them resulting in termination of their function. It can also inhibit the growth and improvement of larvae (Ruuhola, 2001). Phenolics can be regulated for exterior stipulations like light and nutrients; when a plant undergoes stress conditions, it produces lesser phenolics as compared to normal conditions (Cheeke, 1989).

4.4.3. Role of Tannins

Tannins constitute numerous groups of plant secondary metabolites concerned in the induction of plant defense towards insect pests. They have an astringent, sharp and bitter taste that discourages and diverts insect pests away from the host plant. Tannins make hydrogen or covalent bond with the insect mid-gut proteins and digestive enzymes to precipitate the insects. It results in their limited availability to the insect pests and ultimately decreasing the pest population and their subsequent development (War, 2012).

Tannins also can chelate metal ion, leading to the formation of lesions in the mid-gut of insect pests. Deterrence method used for feeding by utilizing the condensed tannins has been examined as studied in a variety of insects and pests, for example, brown tail,

Euproctis chrysorrhoea; and cowpea aphid, *Aphis craccivora*; (Grayer et al.1992). The introduction of tannins production in vegetation as a reaction to insect herbivory and their implementation in insect pest disease control has been thoroughly explained (War, 2012).

For instance, tannin production in *Pinus* spp, *Quercus* spp. and groundnut (Roitto, 2009; War, 2015; Rossi, 2004) showed to exhibit synthesis and induction of tannins upon insect infestation utility of plant defense elicitors.

Insect pests have not only customized produce tannins plant defensive tannins (Zhu-Salzman, 2015), they also utilize them for their growth and developmental process. Though the precise mechanism of insect adaptation to tannins is now not known, greater intestine pH and decrease oxygen levels cease the autoxidation of tannins into poisonous compounds (Johnson, 2000). Moreover, in some species of caterpillars, despite low oxygen amount in the gut, autoxidation of tannins results in the synthesis of toxic products.

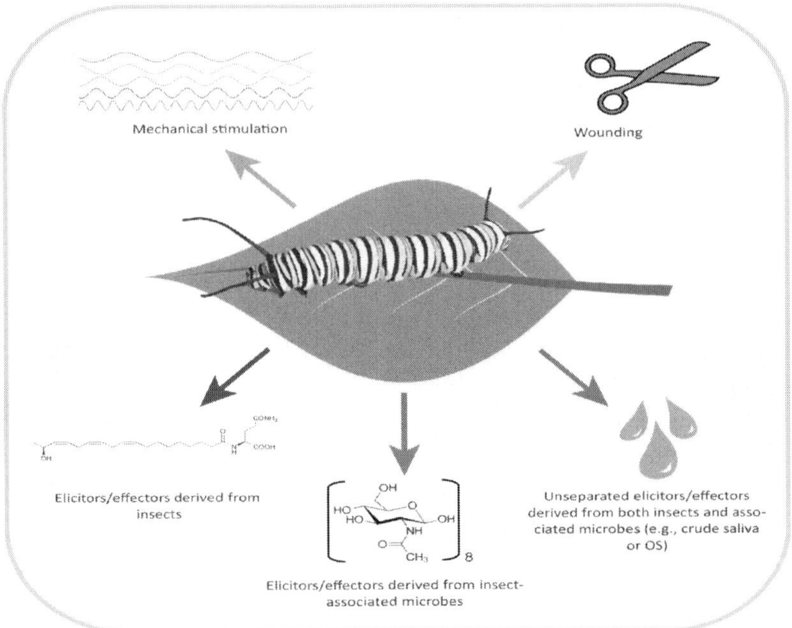

Figure 4.11. Stimulated Herbivory.

Source: (Holopainen et al., 2009) (Jamie M Waterman,,2019)

4.4.4 Cardenolides

Cardenolides are additionally concerned with the induction of defense in opposition to insect herbivores (Green, 2011). Cardenolides normally existing in tolerable levels in plants, they areinduced at the time of insect infestation (Agrawal, 2012). Many pests such as monarch butterfly and crimson milkweed beetleinduced the levels of these compounds in *Asclepias syriaca* (Vannette, 2011). Jasmonic acid transduction routes are triggered by using insect herbivory mediates that assisted by the manufacturing of Cardenolides (War et al., 2018)

Jorgensen, (2003) investigated that sodium Na+ and potassium K+ gradients are incredibly essential for the protection and management of the secondary movement and membrane potentials of Cardenolides in pests and insects.

Cardenolides cease the Na+/K+-ATPase pump necessary for the exits of Na+ out and entry of K+ into the cell. Slight changes in ion gradient may lead to drastic consequences on pest growth and development. Cardenolides for example in the case of digitoxin and cymarin scare off larvae of cabbage looper (Akhtar et al., 2003).

Sachdev-Gupta, (1993) demonstrated that deter action mechanism can also be seen in treacle mustard, *Erysimum cheiranthoides*, *P. rapae* larvae have been also discouraged by utilizing Cardenolides to protect the plant against the harmful effects of the pests. Agarwal, (2004) investigated that the increase in energy use of *Aphis nerii* was once located to show negative correspondence with Cardenolides.

The one biggest drawback of cardenolides is associated with their counter-adaptations mechanism, as they are highly poisonous for insects. It has developed a great concern of their utilization in insect pest control and management (Groen, 2017).

A quantity of elements has been advised to minimize the ability of cardenolides in insect pests. Amino acid replacement in Na/K-ATPase has been seen to be associated with the counter-adaptation of bugs to cardenolides. When changes and substitution of amino acid take place, it jams the cardenolides and decrease their bond formation capacity to the Na/K pump (Dobler, 2015) and amplify the level of K+ ions in insect hemolymph, resulting in abnormalities in cardenolides to inhibits their functioning (War et al., 2018).

The formation of micelles with the aid of the cardenolides was seen in peritrophic membranes of midguts of *S. gregaria* along with *P. americana* makes it impenetrable to cardenolide digitoxin (Barbehenn et al., 1999). *Daphnis nerii* and *Euploea core* larvae's do not include any recognized substitutions in the Na/K-ATPase. Nonetheless feed on plant life with Cardenolides by exhibiting insensitivity closer to these compounds (Agrawal et al., 2015). Other than amino acid replacement in Na/K-ATPase pump, metabolic based detoxifications and efflux

Increased amounts of carbon dioxide from 350–700 ppm prompted the expression of cardenolides in *D. lanata* by way of 60–80%, whilst the water stress decreased the ranges of cardenolides (War et al., 2018). Furthermore, Vannette and Hunter (2011) stated that, both reduction in cardenolides and no impact in response to improved ranges of CO2 in *A. syriaca*.

Moreover, a decrease in the concentration of cardenolides in response to nitrogen fertilization has been reported in *E. cheiranthoides*, the non-living parameters affecting cardenolides manufacturing reveal that environmental factors play an essential function in the insect–plant interaction.

Further, the application of fertilizers on cardenolides manufacturing is a main objective as plants need N: P: K fertilizers for desirable increase and improvement for their growth. Any contrasts or abnormalities in their composition would have a negative impact on the plant's functioning. Thus, similarly extensive studies and investigations on insect adaptation to cardenolides and how fertilizers impact on cardenolide functioning and production are required.

4.5 PHYTOALAXINS

Plants want to reply hastily to herbivory and produce defense chemicals at the website of attack. Plants are constantly working to respond to their pest by activating the synthesis of defensive chemical compounds on the point of attack. One of the most basic naturally occurring techniques is adopted by plants that allow them to recognize the attack and then set off the transcription of defensive genes that produce certain proteins and encodes enzymes to produce defense chemical compounds. Pest generated protective chemical compounds known as phytoalexins can have sizable functional effects, particularly on feeding deterrents.

An early example of induced pest production of phytoalexin was once determined in peas plant (Pisum sativum), where the elicitor Bruchin B has been produced by the pea weevil (Bruchus pisorum). It was shown to set off a cytochrome P450 iso-flavone synthase, leading to the subsequent deposition of the chromanocoumarane pisatin (Cooper, 2005). In maize (*Zea mays*), metabolic analysis of European corn borer (*Ostrinia nubilalis*) feeding tunnels in stems leads to the identification of maize diterpenoid phytoalexins.

Extensive studies on the metabolic analysis of *Zea mays* showed that *Ostrinia nubilalis* forms feeding tunnels in stems of the maize plant. Diterpenoid phytoalexins were isolated from infected maize plant from *Ostrinia nubilalis* commonly known as European corn borer (Hu, 2011). Experimental studies in maize plant showed that prolonged in counter of maize stem by *Ostrinia nubilalis* leads to the plant-based elicitor production and accumulation of nt-kauranediterpenoids such as:

- ent-kauran-17-oic acid
- ent-kauran-17,19-dioic acid
- ent-kauran-19-al-17-oicacid (Schmelz)

Further studies along with chemical extraction found that (a) ent-kaur-15-en-17-oic acid, (b) ent-kaur-15-en-17, (c) 19-dioic acid, and (d) ent-kaur-15-en-19-al-17-oicacid, also named as kauralexins B1 via B3, sequentially.

Detailed quantitative evaluation of European corn borer infested plant tissues showed that four kauralexins namely A1, B1, B2, and B3 were elicited substantially considerable extent than those determined in broken controls.

Surprisingly, kauralexins A2 and A3 gave no contrasting results from damaged controls, suggesting that the pest-based induction of phytoalexins can be precise to non-identical compounds within the identical class of metabolites (Yactayo-Chang et al., 2020). Subsequent studies in maize also proven that European corn borer stem tunneling additionally elicited the deposition of sesquiterpenoid zealexins at the point of injury or wound.

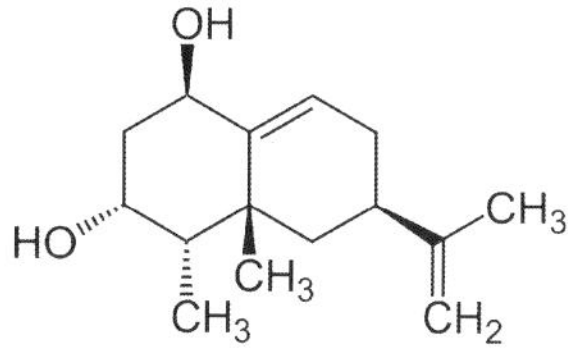

Figure 4.12. Chemical Structure of Phytoalexin.

Source: (Holopainen et al., 2009) (Wikipedia)

Oryza sativa, when attacked by pests, resulted in the production of insect-based phytoalexin that generates a herbivore-specific response. Investigation on *Sogatella furcifera,* common name white backed plant hopper when it attacks, elicits a strong plant hormone reaction that causes the deposition of the diterpenoid momilactone A, whose infestation duration period is approximately 7 days. Furthermore, unstoppable feeding by means of the loreyi leaf worm larvae initiates the elicitor response by actively synthesizing momilactone A and momilactone B in order to generate a defense response (Shinya,2016).

Conversely, feeding by *Spodoptera mauritia* and *Parnara guttata* allow synthesis of low and variable amount of momilactone A and momilactone B, recommended that phytoalexin action to pest attack are continuously changeable and reliant on the precise plant-pest relationship (Yactayo-Chang et al., 2020). A common mode of induction of pest-based production of phytoalexins has also been discovered in soybean (*Glycine max*), inclusive of extremely high levels of the phytochemicals. It majorly comprises of flavones and genistein as a responsive action when aphid attacks on Glycine max. Furthermore, no longer all isoflavones no longer considerably change at some stage in aphid attack on *Glycine max.*

The above-mentioned examples demonstrate the inducible property of crop phytoalexins in response to diverse herbivore attackers' and emphasize their roles as direct protection compounds. One of the most beneficial points of pest-based phytoalexins production is their immediate feeding deterrent action against plant pests. Early studies explained that the phytoalexins such as isoflavins 3R-(-)-vestitol and sativan identified and isolated from a member of *Lotus penduncculatus* had considerable feeding deterrent performance against larvae of *Costelytra zealandica*. Special experimental studies found that pastures with >20% *L. penduncculatus* result in the virtual absence of grass grub beetle, indicating a strong function for pea plant inducible chemicals in insect protection (Cooper, 2005).

Another promptly discovers testing the feeding deterrent venture over lupin (*Lupinus angustifolius*) root-derived flavonoids permanency confirmed up to the expectation through the formation of following phytoalexins:

- Isoflavones
- Isoflavan vestitol
- Maackiain
- Medicarpin,
- Pisatin,
- 20-methoxy phaseollinisoflavin

All above-mentioned chemicals displayed substantial feeding activity in opposition to the grass grub beetle and to the African black beetle. Importantly, partial phytoalexins hold demonstrated bioactivity past agricultural implementations. The soybean phytoalexin glyceollin has shown biomedical importance throughout various

human health areas which include anti-proliferative, anti-oxidative, antiestrogenic and anticholesterolemic capabilities (Phem, 2019). The doable for glyceollin in conformity with have and quintessential role in agricultural purposes also exists. When purified glyceollin was once as soon as applied to *Pharsalus vulgarus that is the common beans* at varying concentrations, strong feeding deterrent activity against the spotted cucumber worm (*Diabrotica undecimpunctata howardi*) or the Mexican kidney bean bloodworm was also determined. In maize, the agronomic characteristic feature concerning phytoalexin bioactivity towards bugs hazel so been demonstrated. An exogenous dose of kauralexins A3 and B3 on the excised surface of maize stems notably reduced European corn borer maize utilization as compared to controls. An additional example in maize was once found in a type of phytoalexins much less in many instances related with plant-based chemical activity, namely oxylipins.

JA, a type of oxylipin is largely recognized for its importance in defense signaling pathways. moreover, a current study tested that the 9-oxylipin and 10-oxo-11-phytoenoicacid, was strongly destruction casing in maize silks and induced a concentration associated inhibition of corn earworm growth (Christensen, 2015) illustrating that numerous phytoalexins can have protective activities against pests and insect attacks (Yactayo-Chang et al., 2020).

4.6. PRODUCTION OF PHYTOANTICIPINS

One of the biggest disadvantages of triggered protection system by plants against pest is the long time period required to produce new compounds. Another strategy is to synthesize the new defensive chemical compounds essentially in the tissues especially which are prone to attack.

Disadvantages of this strategy consist of the metabolic energy required to synthesize compounds even when there is no danger of insect attack, and that the complex structure of compounds are toxic not only to the pests, however additionally to the plant itself.

A common alternate strategy to ignore the poisonous affair is to save the compounds as simply activated non-poisonous types and spark off them at the onset of insect attack.

These compounds are acknowledged as phytoanticipins, as they are synthesized in anticipation of a threat. There are some pathways having a hybrid attitude, though they produce compounds that is integral, although the last operative step is inducible. Maysin is a C-glycosyl flavone that is synthesized essentially at excessive ranges in maize silks and corresponds with multiplied resistance to destructive plant-eating pests inclusive of corn earworm (Yactayo-Chang et al., 2020)

Maysin itself, is not thought to immediately have anti-insect characteristics instead it is actively performing its function by means of transformation into quinones with the aid of polyphenol oxidases that are induced during tissue injury by the insects (Byrne, 1996). Quinones formed from the conversion of maysin and related compounds reduce the accessibility of proteins and associated amino acids as a food source through association to their -SH and -NH2groups (Wiseman, 1995). Different strategies to deal with poisonous compound manufacturing make use of the fact that plant cells are mechanically disturbed during plant-eating pest attacks, which leads to the amalgamation of contents from variable cells and cell compartments.

Phytoanticipins are then removed from their constituent enzymes cells. These chemically compounds are deposited in specialized S-cells. These cells become functional via hydrolysis in the presence of enzyme myrosinases. The myrosinases are stored separately in specialized vegetative vacuoles of specialized myrosin cells. When plant tissues get disrupted by pest attack, leads to the blending of two types of cell contents, and the splitting of the glucosinolates by means of the enzyme myrosinases to generate unstable aglycones.

The aglycones actively repositioned to structure a range of poisonous molecules including:

- Isothiocyanates,
- Thiocyanates,
- Nitriles,
- Epithionitriles (Burow,2017)

Sorghum was seen to synthesize and accumulate the cyanogenic glucoside dhurrin in the cytoplasm, whereas the endogenous -glucosidase, dhurrinase that prompts it is in located and stored in the chloroplasts. Cell disturbance during pest attack, make cyanogenic glucosides disclose to their activating hydrolases, which transform them to cyanohydrin aglycones.

These compounds spontaneously, or with the assistance of enzyme hydroxy nitrile lyases, can be transformed into hydrogen cyanide. A similar strategy is located for the benzoxazinoids located in maize, wheat, and rye that are stored as glucosides in the plant vacuole. These glucosides are hydrolyzed upon tissue disruption by way of plastid localized glucosidases, releasing poisonous aglycones (Frey, 2009).

An alternate method to maintaining the compound's inertness is to synthesize them in specific compartments and excrete them externally in tissues. These are frequently complex compositions of compounds and can be defined as fundamental oils, resins, or latex depending on their manufacturing from glandular trichomes, laticifers, or resin ducts, subsequently (Yactayo-Chang et al., 2020). For instance, glandular trichomes can incorporate a range of defensive metabolites relying on the species and kind of trichome. These can incorporate various complex toxins and terpene volatiles, as nicely as entrapping stiff waxy compounds that act to physically suffocate the insects (Glas, 2012).

4.7. INSECT OVIPOSITION AND PLANT SECONDARY METABOLITES

Insect oviposition is the forward yet essential amongst the chain of events among insect–plant relationships. The fitness of the forces plant vegetation because of fly oviposition determines resistances of pests towards the defensive chemicals. Plants give responses against insect oviposition through prescribe yet triggered inducible defenses, which work in accordance in order to get rid of the grasshopper eggs, accordingly warding off the damage through larvae.

In rice, oviposition with the aid of a white-backed planthopper, *Sogatella furcifera* induces the production of ovicidal composite benzyl benzoate (Yamasaki et al. 2003). The tissue wounding caused by European cranberry bush, *Viburnum* species respond in accordance with viburnum blade beetle *Pyrrhalta viburni* oviposition is a profound protecting explanation to that amount reasons the bad effects on eggs and/or their expulsion. Jasmonates are considered as vital elicitors about oviposition-induced hindrance yet hold eggs over a variety of lepidopteran bugs into higher concentration than between inter tissues yet larval diets (War et al. 2013).

Furthermore, JA-treated plants obtain less number concerning eggs from P. brassicae, P. rapae and H. Armigera when compared to the untreated control plants (War et al. 2013).Oviposition by means of fly pests has been observed to induce genes associated in conformity with SA, which is the main mediator regarding plant defense against pathogens then sap -sucking insects (Zhao et al. 2009). The SA then JA signaling pathways job antagonistically) thus, activation about SA path within plants by means of grasshopper oviposition could result in the suppression of JA signaling transduction path or the low degree defense towards chewing insects (Sánchez-Sanchez, Morquecho-contreras and Morquecho-contreras, no date).

4.8. INSECT RESPONSIVE TOWARDS PLANT CHEMICAL BASED DEFENSE

The evolutionary strain for plant life to continue to improve new and improved defense chemicals is managed with the aid of the capability of bugs to progress to remain safe or keep away from these defenses. Various bugs have progressed to developed specialized techniques to get over their host plant's protection chemicals.

These strategies consist of suppressing host defense responses via the use of products and release special effectors proteins that try to avoid, detoxify or try to become ineffective host defense chemicals. These strategies have now not solely allowed the insects to get their food from their hosts; they have also allowed some insects to undertake the protection systems of their host plants.

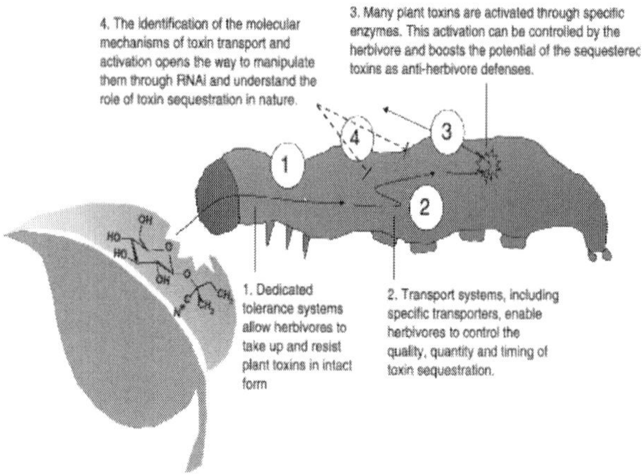

Figure 4.13. Sequestration of Plant Secondary Metabolites by Plant herbivores.

Source: (Matthias Erb, 2016)

An instance of this can be seen in glucosinolate producing brassica plants such cultivars of *Brassica oleracea*. It also includes broccoli and cabbage plants. Brassica specific insects, such as the larvae of the diamond back moth and the wilderness locust, possess sulfatases in their gut, which make them able to get their feed from glucosinolate plants. These sulfatases demobilized glucosinolates by transforming them into an inactive structure that does now not react with myrosinases, permitting the bugs to feed safely.

White butterflies of cabbage plants named, *Pieris rapae* are also specialist herbivores on plants carrying glucosinolate. Their larvae are uncovered to cyanide during the metabolism of the breakdown of benzylglucosinolate product phenylacetonitrile. The larvae have –cyan alanine synthases that allow them to metabolize cyanide to cyan alanine and thiocyanate, for that reason preventing its deposition and toxicity (Van, 2016).

Another strategy is to keep away from the phone rupture that is essential for the stimulation of these compounds. As pore making and then plant sap sucking herbivores, cabbage aphids (*Brevicoryne brassicae*) can avoid the blend of glucosinolates and myrosinases because of their mode of feeding, in order to prevent the damaging of many cells in one attempt. Moreover, these aphids also can seize and capture their host plant's glucosinolates completely intact. This, in addition to the truth, is that aphids contain their own microbodies that carry myrosinase, which permits the aphids to effectively replicate the chemical protection machine of their host plant life (Sánchez-sánchez, Morquecho-contreras and Morquecho-contreras, no date).

Larvae of moths of the genus Zygaena, evolved to have advanced specializations in cyanogenic plants, that can manufacture their very own cyanogenic glucosides, but have additionally verified the capacity to capture the compounds intact from their host plant life. Larvae feeding on cyanogenic plants were identified to develop faster, most likely due to the metabolic benefit of no longer having to synthesize their own cyanogenic glucosides compounds (Zagrobelny, et al., 2006).

4.9. CONCLUSION

The conclusion that can be drawn from the chapter is that the plants usually are evolving and working for generating new strategies to avoid insect pests and insects. In turn, plants are always equipped to improve mechanisms of counter-adaptations. This complex interaction has led to the development of a range of plant shielding qualities and the counter - adaptive aspects in plant-eating pests as well.

REFERENCES

1. Holopainen, J. K. *et al.*(2009) 'Potential for the Use of Exogenous Chemical Elicitors in Disease and Insect Pest Management of Conifer Seedling Production', pp. 17–24.
2. Sánchez-sánchez, H., Morquecho-contreras, A. and Morquecho-contreras, A. (no date) 'Chemical Plant Defense Against Herbivores', pp. 3–28.
3. War, A. R. *et al.* (2018) 'Review SPECIAL ISSUE : Using Non-Model Systems to Explore Plant Pollinator and Plant-Herbivore Interactions Plant defence against herbivory and insect adaptations', pp. 1–19. doi: 10.1093/aobpla/ply037.
4. Yactayo-Chang, J. P. *et al.* (2020) 'Plant defense chemicals against insect pests', *Agronomy*, 10(8). doi: 10.3390/agronomy10081156.
5. Agrawal AA. 2002. Herbivory and maternal effects: mechanisms and consequences of transgenerational induced plant resistance. *Ecology* 83:3408–3415.
6. Ahn JE, Lovingshimer MR, Salzman RA, Presnail JK, Lu AL, Koiwa H,Zhu-Salzman K. 2007. Cowpea bruchid *Callosobruchus maculatus*counteracts dietary protease inhibitors by modulating propeptidesof major digestive enzymes. *Insect Molecular Biology*16:295–304.
7. Sharma HC, Sujana G, Rao DM. 2009. Morphological and chemical components of resistance to pod borer, *Helicoverpa armigera* in wild relatives of pigeonpea. *Arthropod-Plant Interactions*3:151–161.
8. Holopainen JK, Blande JD. 2013. Where do herbivore-induced plant volatiles go? *Frontiers in Plant Science* 4:185.
9. Kaur R, Gupta AK, Taggar GK. 2015. Induced resistance by oxidative shifts in pigeonpea (*Cajanus cajan*L.) following *Helicoverpaarmigera* (Hübner) herbivory. *Pest Management Science* 71:770–782.
10. Karban R. 2011. The ecology and evolution of induced resistance against herbivores. *Functional Ecology* 25:339–347.
11. War AR, Paulraj MG, Ahmad T, Buhroo AA, Hussain B, Ignacimuthu S,Sharma HC. 2012. Mechanisms of plant defense against insect herbivores. *Plant Signaling & Behavior* 7:1306–1320.
12. War AR, Hussain B, Sharma HC. 2013. Induced resistance in groundnut by jasmonic acid and salicylic acid through alteration

13. of trichome density and oviposition by *Helicoverpa armigera* (Hubner) (Lepidoptera: Noctuidae). *AOB Plants*5:plt053;doi:10.1093/aobpla/plt053.

14. Zhao LY, Chen JL, Cheng DF, Sun JR, Liu Y, Tian Z. 2009. Biochemical and molecular characterizations of *Sitobion avenae*-induced wheat defense responses. *Crop Protection* 28:435–442.

15. Huffaker A, Pearce G, Veyrat N. 2012. Plant elicitor peptides are conserved signals regulating direct and indirect antiherbivore defense. Proc Natl Acad Sci U S A.; 110: 5707–5712. DOI: 10.1073/pnas.1214668110.

16. Maffei ME, Arimura G, Mithöfer A. 2012. Natural elicitors, effectors and modulators of plant

17. responses.;29: 1288–1303.

18. Maffei ME, Mithöfer A, Boland W.2007. Before gene expression: early events in plant-insect

19. interaction. Trends Plant Sci.; 12: 310–316.

20. Pieterse CMJ, Van der Does D, Zamioudis C.2012. Hormonal modulation of plant immunity. Annu Rev Cell Dev Biol.; 28: 489–521.

21. Consales F, Schweizer F, Erb M. 2012. Insect oral secretions suppress wound-induced responses in *Arabidopsis*. *Journal of*

22. *Experimental Botany* 63:727–737.

23. Wu J, Hettenhausen C, Meldau S, Baldwin IT. 2007. Herbivory rapidly activates MAPK signaling in attacked and unattacked leaf regions but not between leaves of *Nicotiana attenuata*. *ThePlant Cell* 19:1096–1122.

24. Schoeneweiss, D. F. 1975. Predisposi-tion, stress, and plant disease. Ann. Rev. Phytopathol. 13:193-211

25. Painter, R. H. 1951. Insect Resistance in Crop Plants New York: MacMillan

26. Martin, J. T., Juniper, B. E. 1970. The Cuticles of Plants. New York: St. Martins

27. Beck, S. D. 1965. Resistance of plants to insects. Ann. Rev. Entomol 10:207-32

28. Stolman, A. 1967. Combined action of drugs, with toxicological implications. Progr. Chem. Toxicol. 3:305-61

29. House, H. L. 1961. Insect nutrition. Ann. Rev. Entomol. 6:13-26
30. Culvenor, C. C. J. 1970. Toxic plants-a reevaluation. Search 1: 103-110 62. Curtis, J. D., Lersten, N. R. 1974. Mor-phology, seasonal variation and func-tion of resin glands on buds and leaves of Populus deltoides (Salicaceae). Am. J. Bot. 61:835-45
31. Cooper, L.D.; Doss, R.P.; Price, R.; E.2005. Application of Bruchin B to pea pods results in the up-regulation of CYP93C18, a putative isoflavone synthase gene, and an increase in the level of pisatin, an isoflavone phytoalexin., 56, 1229–1237.
32. Hu_aker, A.; Kaplan, F.; Vaughan. 2011.Novel acidic sesquiterpenoids constitute a dominant class of pathogen-induced phytoalexins in maize. Plant Physiol. 2011, 156, 2082–2097.
33. Shinya at el., 2016. Modulation of plant defense responses to herbivores by simultaneous recognition of different herbivore-associated elicitors in rice., 6, 32537. [CrossRef] [PubMed]
34. Pham, T.H.; Lecomte, S.; Efstathiou.2019. An update on the effects of glyceollins on human health: Possible anticancer effects and underlying mechanisms. Nutrients , 11, 79. [CrossRef]
35. Christensen, S.A.; Huffaker, A.; Kaplan.2015. Maize death acids, 9-lipoxygenase-derived cyclopente(a)nones, display activity as cytotoxic phytoalexins and transcriptional mediators. Proc. Natl. Acad. Sci. USA
36. Zagrobelny, at el., 2006

Chapter 5

RNA and Molecular RNA (miRNA) Role and Mode of Action in Plant-Pest Interactions

CONTENTS

in these organisms. It has been well established in this context that the RNA chaperone Hfq is involved in several RNA-based regulatory processes in prokaryotes, and that it plays a key role in these systems.

Ago proteins from bacteria have also been implicated in several other types of RNA-influenced gene regulation processes (Watkins et al., 2019). Furthermore, the CRISPR-CAS system, which is made up of clustered regularly interspaced short palindromic repeats and related genes, has gotten a lot of attention because of its exceptional ability to perform RNA-guided genetic modification. In a strange twist of fate, this method was responsible for the award of the Nobel Prize in Chemistry for the year 2020. Even though prokaryotic regulating RNAs differ from their sophisticated eukaryotic counterparts in several aspects, including size range and associated pathways, their importance to regulatory systems in a variety of activities has been demonstrated(Dutta,2018).

A fresh set of studies concerning the role of extracellular regulatory RNAs in cell-to-cell contacts is creating excitement, even though most research on regulatory RNAs and their pathways has focused on the intracellular level, for the most part. There are several examples of interspecies and even cross-kingdom RNA-based communication, notably those involving host–parasite/symbiont interactions that have been documented.

Insects also have a substantial impact on the environment. This has resulted in various recognized modes of communication between insects and their associated host plants, which include specific protein kinases, volatiles, and other compounds. This has resulted in various recognized modes of communication between insects and their associated host plants.

It is becoming increasingly clear that insects, as well as plant–insect interactions, are capable of cell-cell communication that is RNA-dependent. Discovering the kinetics and mechanisms of intercellular RNA transmission in insects, as well as the connections between members of the plant kingdom, has the potential to help in the development of new technologies.

When it comes to dealing with insect-related difficulties, RNA-directed pest control approaches are essential, and they are a crucial part of the solution (Santos *et al.*, 2021). Plants are an essential component of the environment, and they can be utilized by people for a variety of purposes, including housing, medicine, and food consumption. Human activities, non-living stresses such as drought, poor soil conditions, environmental degradation, and living threats such as insects, pests, and bacterial diseases, to name a few, all have a significant impact on plant production.

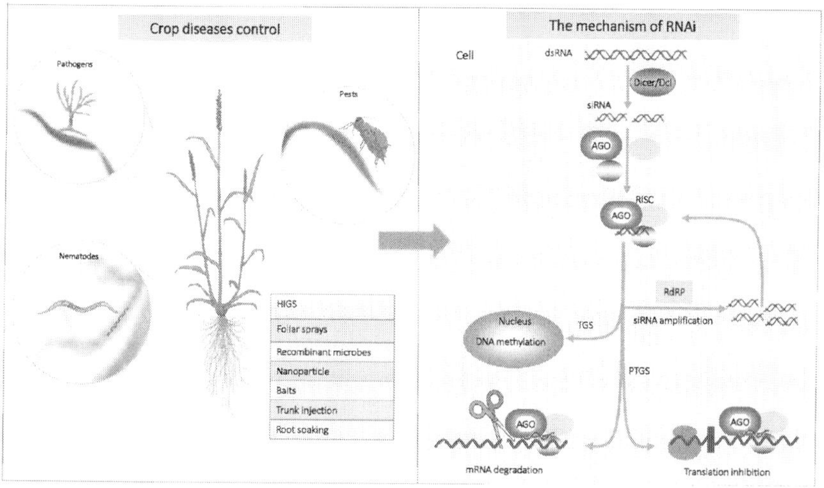

Figure 5.1. Role of RNAi protecting the Wheat Plant against pests and other Insects.

Source:(Liu et al., 2021)

The rise in the urban population, along with the corresponding increase in consumption of wild plants, justifies an increase in agricultural yields in the coming years. Among these methods is the development and implementation of environmentally sustainable pest control techniques, which can help to reduce production implications.

MicroRNAs (miRNAs) have been discovered to have a role in the regulation of important plant metabolic activities at the post-transcriptional level, thanks to recent advances in molecular biology (Bej et al., 2014). Plant productivity can be improved by manipulating these RNA regulators, which might be useful as potential compounds for improving plant production.

5.2. MIRNA BIOSYNTHESIS AND OCCURRENCE IN PLANTS

According to multiple studies, various microRNAs have been found to be evolutionarily conserved throughout all major plant groups, including angiosperms, ferns, and gymnosperms, as well as across all other plant species, including fungi (Sun et al., 2012). The preservation and diversity of miRNAs over the history of the plant world may be classified into two categories, which can be further divided into subcategories.

Prehistoric miRNAs, on the other hand, are abundantly conveyed and have evolved to be conserved, whereas early miRNAs are expressed at significantly lower rates and are thus triggered under special circumstances, and as a result, they are typically found within only a few organisms, as a result of which they have been preserved from an evolutionary standpoint (Qin et al., 2014). The existence of miRNA genes in Arabidopsis plants, as well as indications of recurrent entry and removal have recently been discovered in the plant. Moreover, it demonstrated that a huge number of microRNAs (miRNAs) are often disrupted throughout evolutionary processes (Fahlgren et al., 2007).

In addition to being non-coding RNA molecules, miRNAs have a range of sizes ranging from 20 to 24 nucleotides in length (Jones et al., 2006). Introns in the antisense (or sense) configuration of MIR genes (also known as microRNA genes) are usually found in intergenic regions, but they may also be found in introns in the antisense or sense configuration of MIR genes. Numerous microRNAs (miRNAs) are found in the genome in association with one another as long polycistronic RNAs.

MicroRNAs (miRNAs) are generated in the nucleus by RNA polymerase II from a particular MIR gene. A long main transcript is produced, which is subsequently spliced to produce mature miRNAs, all of which are present in the nucleus (Jones et al., 2006). Typical transcripts have their 5′ ends capped and their 3′ ends polyadenylated, and this is one of such transcripts (Pantaleo et al., 2010).

PREMIRNAS are generated when pri-miRNAs are cleaved by enzymes that look like RNaseIII, such as DICER-LIKE (DCL1), in conjunction with a variety of other proteins, including HYL1 and SE (pre-miRNAs). DCL1 is responsible for the formation of miRNA:miRNA duplexes, which are then exported to the nucleus instead of being processed in the cytoplasm.

HUA enhancer 1, which contains the S-adenosyl-L-methionine-dependent RNA methyltransferase, methylates each duplex, and then EXPORTIN-5 (HASTY), which exports the hydroxyl group to the cytoplasm (Park et al., 2005), is involved in the methylation of each duplex. The RNA-induced silencing complex (RISC), also known as RISC-AGO, is where the duplexes are loaded into the complex.

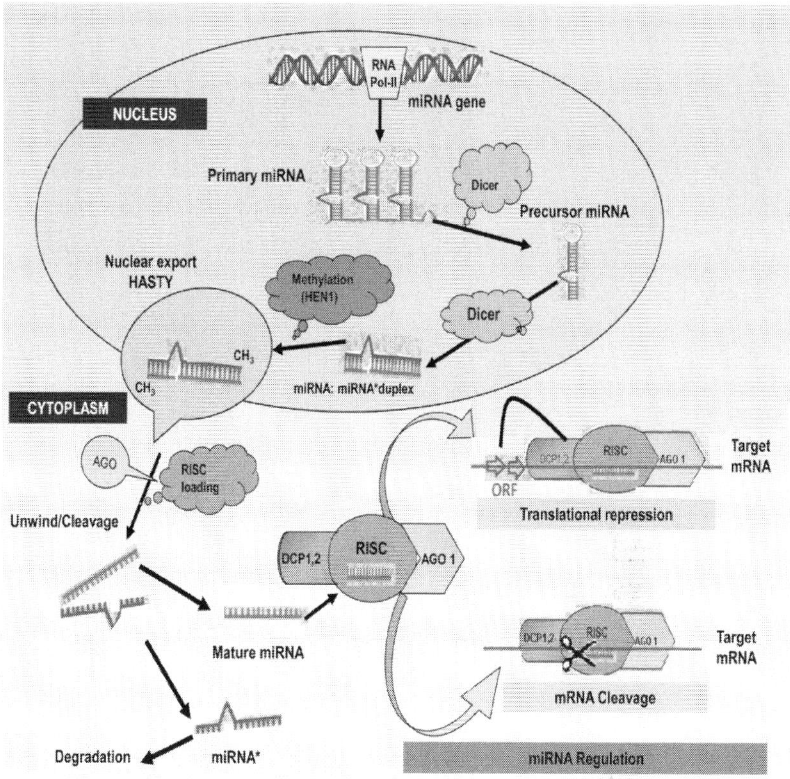

Figure 5.2. Biosynthesis of miRNA in Plants.

Source: (Djami-Tchatchou et al., 2017)

The AGO1 protein (Djami-Tchatchou et al., 2017) aids in the unwinding of miRNA: miRNA duplexes in part by directing one strand of the duplex miRNA to the exosome, where the mature miRNA is destroyed by a tiny RNA degrading nuclease.

In addition to this, the maturing microRNA (miRNA), the RISC, is directed to the corresponding target mRNAs by the miRNA. Translational suppression and/or mRNA cleavage are caused by microRNAs (miRNAs) that have perfect base matching to their target RNA (Sun et al., 2012).

5.3. REGULATION OF GENE EXPRESSION OF MIRNA

Although there are numerous tiny RNAs, their variety is adequate to identify virtually any possible RNA, therefore governing essentially every element of plant science, which, in turn, influences transcription factor regulation (TFs).

The mechanisms of working of miRNAs as significant gene regulation regulators have been intensively utilized (Djami-Tchatchou and Dubery, 2015). Typically plant miRNAs are between 20 and 24 nucleotides long, and they affect gene expression largely by down regulating it after protein synthesis has occurred. This is done via a greater extent of compatibility.

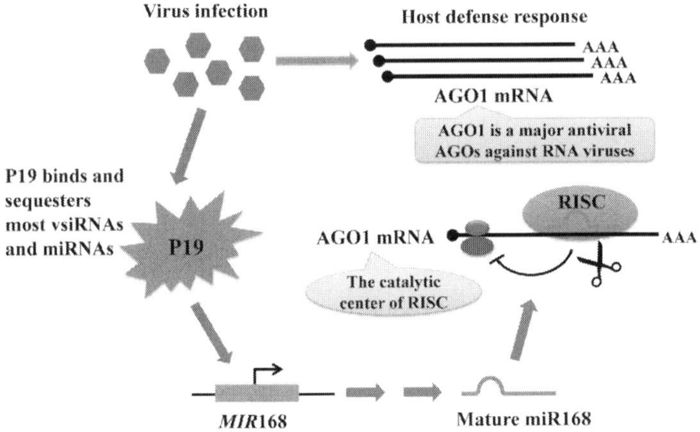

Figure 5.3. Plant miRNA action Mechanism.

Source: Sheng-Rui Liu, (2017)

Poly (A) tail deletion might cause aimed at mRNAs to be destabilized, contributing to their mRNA degradation (Guleria et al., 2011). miRNAs could assist give resilience to biochemical functions by serving as post-transcriptional inhibitors of expression of genes, as well as helping to silence random variations in transcript expression levels. Heterogeneous-length miRNAs could be synthesized within the same MIR gene, and heterogeneous-length miRNAs can be created from different molecular RNA genomes (Djami-Tchatchou et al., 2017).

DCL1 is involved in the production of most plant-based pre-miRNAs, while DCLs 2–4 also capable of processing additional lengths of mature miRNAs. This miRNA variability increases the regulatory capability of the miRNA populations. Additionally, with this kind of miRNA heterogeneity, synthetic miRNA expression cassettes capable of releasing patterns of artificial miRNAs may be developed.

5.4. MOLECULAR APPROACHES TO STUDY MIRNA

To understand the activities of microRNAs, it is necessary to first identify them. Micro RNAs have largely been identified using methods such as molecular cloning and sequencing, genetic screening, and bioinformatic estimations (Schommer et al., 2012). Short RNA pools were isolated, cloned, and sequenced in *Arabidopsis thaliana*, resulting in the discovery of the very first plant miRNAs such as

- miR156,
- miR159,
- miR164, and others (Llave et al., 2002).

A genomic screen hasn't turned up as many miRNAs as we may have expected due to the smaller size of the short RNAs and overlapping with other miRNA-coding genes. Nevertheless, in Arabidopsis, utilizing the up - regulation technique and genetic testing, it has been shown that the activation-tagged method for miRNA identification is efficient at isolating predominant miRNA mutants, including such as miR172a. The search for new traits led to the identification of the JAW (jagged and wavy) locus, which generates microRNA (miR-JAW) that can assist in the cutting of TCP gene mRNAs (TFs) responsible for leaf growth (Palatnik et al., 2003).

Nucleotide mutations with insertions/deletions in tandem with the loss of miRNA binding sites would lead to gains or losses of miRNA target sites during co-evolution of miRNAs and their gene targets, and screening for these mutations may help identify developmental abnormalities. Analytical and experimental approaches have been employed to find countless miRNAs in vegetation during the last decade.

Sequence complementarity is strong among plant miRNA targets, and homolog-based comparative genomic methods predominated in the early computational studies. miRNAs were discovered using the BLAST algorithm on the known miRNA sequences to compare with expressed sequence tags (ESTs), genome sequence surveys, and genome sequences of various plant species.

Genetic screening methods and methods of direct cloning are among the first of the early experimental techniques used in identifying and functionally analyzing miRNAs within crops. The sequencing technique has been introduced to be a more effective instrument for miRNA identification and target identification in plants by a series of advances: molecular cloning and Sanger sequencing. miRNAs discovered by methods such as Northern blotting or extremely semi-quantitative real-time PCR tests must be verified for translation (Zhang and Wang, 2015).

Furthermore, many techniques such as:

- Protein chromatography,
- Mass spectrometry,
- Blotting techniques, Protein foot printing, and other techniques can be combined to identify miRNA targets at the target sites.

The latest research employing cell cultures, for example, examined the impact of miRNA modulation on the levels of numerous proteins using a quantitative mass spectrometry-based method which is based on SILAC, which is stable isotope labeling with amino acids in cell culture along with microarray result management (Baek et al., 2008). Their findings revealed this. In most relationships, miRNAs function as rheostats to fine-tune protein production. The discovery of miRNAs resulted in the creation of databases offering accessible knowledge regarding miRNAs. Important databases are as follow:

- http://www.mirbase.org is the most valuable tool for miRNA investigation since it is a complete and accessible miRNA library based on miRNA identity, keyword, references, and interpretation (Zhang and Wang, 2015).
- The Plant MicroRNA Database (PMRD; http://mirnablog.com/plant-microRNA-database-goes-online/) provides detailed information about plant miRNAs, including the miRNA and its target, secondary structure, diagnostic conditions, as well as a genome browser (Zhang et al., 2010).
- Powerful mathematical algorithms such as Next-Gen sequence databases (https://mpss.udel.edu/index.php), AthaMap (http://www.athamap.de/), and indeed the CLC Genomic workbench 6 software were also used to help with the NGS processing (CLC Bio, Cambridge, MA, USA).

5.5. COMMUNICATION METHOD THROUGH MIRNA

Overwhelming evidence supports the idea that RNA molecules are present in extracellular settings, primarily as sRNAs. Since RNA molecules were discovered to be involved in interspecies communication, it has become apparent that these molecules function as a communication layer even between distantly related species (Zeng et al., 2019).

In plants, mobile RNAs have a well-established function in intercellular communication (Ham, et al., 2017). Extracellular RNAs have been widely studied in mammalian biofluids, especially in mammals. Once secreted, these molecules have been seen to be taken up by recipient cells, which is a normal cell process. Arthropods and nematodes, only with a restricted number of experiments proving functional RNA transmission, show extracellular RNAs in invertebrates.

Exocellular RNAs are currently the subject of a lot of research, with the possible exception being their involvement in intercellular/intraindividual communication, which has been studied in insects like beetles and flies. siRNA communities were noted throughout the hemolymph of the Korean rhinoceros beetle, *Allomyrina dichotoma*; miRNAs have been found in the cell-free hemolymph of Drosophila melanogaster flies, and some have been seen to exhibit age-dependent expression; multiple extracellular sRNA communities have been found in the D. melanogaster cell lines S2R+ and D17-c3; sRNA groups have also been discovered in the extracellular media of *D. melanogaster* molecules, namely in S2 and Cl8 cell lines, with indications of differential secretion; siRNA populations have been discovered in the hemolymph of Drosophila flies, and have been demonstrated to spread the RNAi signal (Liu *et al.*, 2021). Beyond their role in cell-to-cell communication inside single species, RNA molecules can mediate connections across different individuals, according to some researchers. Interindividual RNA-based transmission has been discovered in social animals, which means this phenomenon is researched mainly in insects. This study focuses mostly on the honey bee, *Apis mellifera*, where it is being determined that nurse bee secretions have a bearing on caste determination. Recently, an RNA route has been established that transmits information between generations, between the hemolymph and then in the jellies (Santos *et al.*, 2021)

5.5.1. Communication Method through MiRNA between Species and Kingdoms

In addition, transmission of operational RNA across species has also been documented in animals. Another way to generically call interspecies RNA mediated interaction is called interspecies RNA dependent communication, and the only known examples thus far are host–parasite relationships. This is an intriguing case that takes place in the world of insects.

Cotesia vestalis as well as its host, the diamondback moth, are both involved. Whereas the growing wasp larva does have the potential to discharge teratocyte cells into another host body, it does not do so yet. As these cells produce extracellular vesicles, they release the miRNA Cve-miR-281 into the surrounding media (Santos *et al.*, 2021). The larval wasp of *C. vestalis* species also produces a *C. vestalis bracovirus* ,which causes the Cve-miR-novel22 miRNA to be expressed inside the host.

It turns out that the host's ecdysone cascade is controlled by both miRNAs, which delays the growth of the parasite, therefore giving the parasite time to grow (White et al., 2020). Additionally, it has been found that these compounds are present in mosquito saliva and other blood-feeding arthropods, leading to conjecture regarding their involvement in regulating the host's response. Comparable studies have demonstrated that parasitic nematodes produce RNA molecules that control host biology (Buck et al., 2014).

The notability of interspecies RNA communication within plants has also been confirmed. The data provided for parasitic plants in the genus *Cuscuta* and associated hosts is based largely on bidirectional RNA transfer, as proven in a study. Additionally, miRNAs isolated from *Cuscuta* have been shown to target genes in Arabidopsis.

A final factor in the soybean *Glycine max-Phaseolus vulgaris* grafting structure is likely to be the transport of short RNAs between stems to the root in a unidirectional movement. An instance of inter-kingdom RNA-mediated communication between plants or pests and symbiotic microbes has also been documented. Ribonucleic acid transmission has been observed in both fungi and protests across plants, animals, and Fungi and Protista (Santos *et al.*, 2021).

5.5.2. RNA Based Interactions of Plants and Pests

There are only a few papers have been published so far on RNA transfer from insects to plants, indicating that little is known about this process. The green peach aphid, *Myzus persicae*, revealed the first hint of functional RNA transmission between plants and insects. In this work, a reduction in aphid fertility was seen in insects that relied on Arabidopsis mutants that negatively impact the miRNA system.

This was taken as evidence that plant miRNA expression may contribute to aphid infestation. Subsequently, it was discovered that sRNAs produced by *Bemisia tabaci*, the whitefly that feeds on the phloem of plants, may be found in the tissue and phloem of the tomato plant. Interestingly, in the most recent research, aphids infected with the pest *M. persicae* were discovered to transmit the Ya transcript to distant leaves (Santos *et al.*, 2021).

This transcript was shown to be a virulence factor of long noncoding RNA, and it was proven to induce aphid fecundity (Zhang et al., 2012). Many researchers disagree about the abundance and usefulness of dietary RNAs within animals over the past century (Masood et al., 2016). Although growing functional evidence suggests that the presence of certain insect-plant RNA-based systems leads to implementation, decreasing evidence points away from such a system in insects.

While there are several studies that detected plant-based RNAs in herbivorous insects, and some even show activity, all agree that these RNAs serve a functional purpose. Stable plant miRNAs have been seen in the aphid bug, *Aphis gossypii*, and the presence of these miRNAs is demonstrated by the existence of *Cucumis melo* miRNAs.

Silkworm mRNAs were discovered in both the hemolymph and the tissues of the silkworm, Bombyx mori. Sorghum miRNAs, previously found in the greenbug, *Schizaphis graminum*, as well as the yellow sugarcane aphid, *Sipha flava*, were later detected in a different aphid species, the yellow sugarcane aphid, *Sipha flava*. These miRNAs have been implicated in *in silico* studies of aphid fecundity, implying a role in aphid infestation.

This study discovered miRNAs in the stomach of the green peach aphid, *M. persicae* that belongs to the *Brassica oleracea*. The predicted miRNA targets in the insect were mostly genes involved in carbohydrate metabolism, modification, and nuclear structure. Other studies have revealed that plant-derived miRNAs are found in the hemolymph of the diamondback moth, *P. xylostella*.

Additionally, two of the most prevalent plant miRNAs discovered were confirmed to target a class of genes known to be critical to arthropod hemolymph composition. Besides this, it was found that miRNA for plants that are quite numerous, such as miRNA found in wheat, barley, and rice, are known to impact the rate of pupation and egg hatching (Zhang et al., 2019).

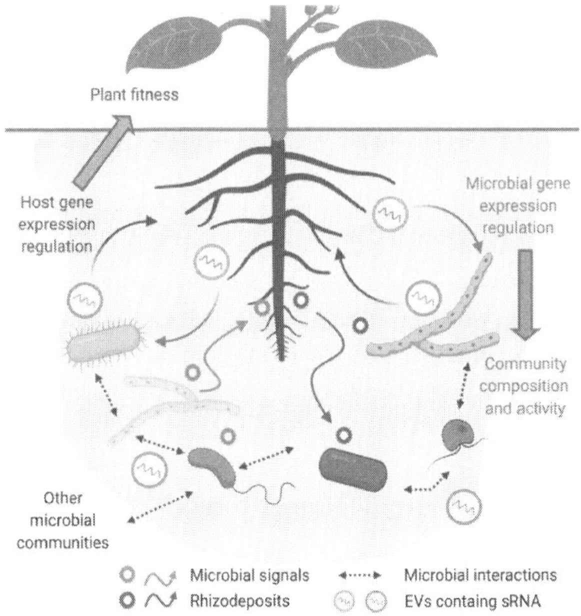

Figure 5.4. Plant miRNA action Mechanism.

Source: Harriet Middleton, (2020)

Many plant miRNAs that are present in the honeybee's larval diet contribute to determining the caste of the bee. Many of the miRNAs found in beebread are enhanced relative to royal jelly, a foodstuff known to help inhibit the formation of queens and instead encourage the growth of worker bees. Worker bee traits, with comparable findings in D. melanogaster larvae, were produced by feeding miR162a to larvae. *A. mellifera* is a crucial factor in developmental cast formation (Zhu et al., 2017)

5.5.3. Artificial Plant-to-Insect RNA Transfer

Finally, when plant-insect functional transfer of RNA is engineered, it is essential that we may study plant-insect functional transfer of RNA. In essence, genetic modification of host plants may be used to direct RNA molecules to certain insect genes. When the insect consumes the plants, these RNA Nucleotides enter the insect body, and then they reach the cells of the recipients, where they trigger the inhibition of gene expression via RNAi.

Some transgenic plants engineered to produce an RNAi response in insects are known as host-induced gene silencing. The first research showing that transgenic plants are capable of resisting insect attacks that employed dsRNA began in 2007 (Mao et al., 2011). In the six decades since the original study, several further proofs have been published.

There are instances in which lengthy dsRNA strands are expressed. Because of this, there is a strong possibility that insect miRNAs may be expressed in plants as well. Furthermore, miRNA may be included in plants or sprayed on plants, and the plant-expressed insect miRNA precursor was found to be more efficient than injecting the miRNA directly (Belly et al., 2020).

Based on the absence of Drosha, a key nuclear processing enzyme for insect miRNA synthesis, the authors propose that insect-based miRNA precursors fail to be processed by the plant.

To generate RNIs, a molecule must be placed in the chloroplast rather than the cellular cytoplasm, involving chloroplast genome alteration. Since the organelle contains no RNAi processing machinery, plant cells can concentrate more RNA molecules, which increases the effectiveness of the RNAi response in insects (He t al., 2020). Despite this, it is still an inefficient strategy, as RNA molecules that originate in the chloroplast are only seldom passed over to the insects.

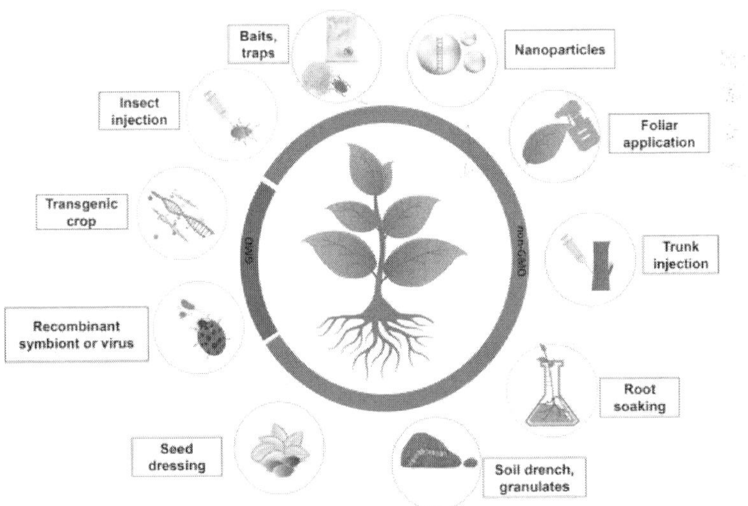

Figure 5.5. RNA-based technologies for pest control in plants.

Source:Image by Shaoshuai Liu, 2020

Since there are RNAi-based pest-resistant plants accessible, the very first RNAi-based pest-resistant plant has appeared on the market. The product was created by Dow Agrosciences and Monsanto, and was called SmartStax Pro. The preceding image highlights the significance of this area in the economic sector, as well as the imperative for researchers to understand the plant-insect RNA transfer processes, which may lead to the creation of more efficient pesticides, which use artificial RNA (Santos et al., 2021).

5.6. METHOD OF RNA TRANSFER

It has been reported that extracellular RNA transports information to insects. Plant-insect RNA transfer has also been documented, and, in certain circumstances, the RNA signal has been found to be functional. Using plant-insect relationships as well as genetically modified plants, scientists were able to show the ability to target insects using RNA-based pest management. Still, very little is understood about the ways mobile RNAs are transferred across insect cells, or in insects. In addition to these three established cell-to-cell RNA communication methods, which use naked RNA, RNA bound to RBPs, and RNA-containing extracellular vesicles, there are still three emerging strategies that can be applied, two of which use RNA and one of which uses the RNA present in extracellular vesicles (EVs) and are as follows:

- Naked RNA
- RNA and RNA Binding Proteins
- Extracellular Vesicles Carrying RNA

5.6.1. RNA is Naked

Naked RNAs, that is, RNAs that are not encased in endosymbiotic Vps or secured by RNA binding proteins, may be transported over considerable distances in plants (Wang et al., 2020). Plasmodesmata and the desmotubule link the endoplasmic reticulum of two neighboring cells to transmit naked sRNAs. In other words, bare sRNA may rapidly travel from cell to cell across the septal hole (Nazim Uddin et al., 2013). Transfer between neighboring cells in mammals can be achieved through gap junctions. Additionally, it has been reported that siRNA and miRNA-mediated transfer is mediated by gap junctions (Thuringer et al., 2016). There is no research that we are aware of that has investigated the use of gap junctions to transport RNA molecules from one cell to another. Although SID proteins function as channels for intercellular RNA transport in *C. elegans*, transmembrane systemic RNA interference (T RNAi) is deficient. It was hypothesized that short-range movement of dsRNA and RISC components occurs in Drosophila cells via nanotube-like structures.

However, until this hypothesis is proven, it is unclear if these structures will successfully transport naked RNAs across neighboring insect cells. These experiments show that the RNA transport systems in plants, fungi, and mammals could be quite similar. In both plants and fungi, bare sRNA may travel across the apoplast via apoplastic movement. When produced naked sRNAs travel along the cell wall, they can be taken up by other cells and penetrate both nearby and distant compartments (Zhang et al., 2019).

Plants use the vascular system to transport naked sRNAs to distant cells, and naked sRNAs may be found in the phloem as well. In addition, it is notable that numerous findings show bidirectional interkingdom RNAi, that is, RNA interference, occurs across plants and fungi. Using specialized infection structures known as haustoria, fungi and parasitic plants may facilitate the delivery of sRNA to the host and pathogens at the plant-plant and plant-fungi levels.

Naked extracellular RNAs in human plasma are quickly destroyed. Citing previous findings, it may be deduced that bare RNA molecules are speedily destroyed in insect biofluids. However, it has recently been shown that animal extracellular fluids include RNA molecules that remain stable. These three pieces of information collectively lead to the conclusion that in animal body fluids, mobile RNAs must be protected from degradation in order to be effectively transported (Santos *et al.*, 2021).

5.6.2. RNA-binding proteins

A mycorrhizal fungus establishes a receptor-binding protein (RBP) to facilitate both short as well as long-range RNA transfer in plants. Plasmodesmata (aka vascular bundles) and the phloem serve as transport conduits for the small RNA-binding protein *Cucurbita maxima* Phloem Small RNA-Binding Protein 1, which binds and transfers sRNAs between cells via these channels. Other RBPs have also been discovered in various plant phloem.

While Ago proteins have also been proposed to play a role in plant sRNA transfer, they are not directly involved. RBPs are established in plants and are responsible for mediating both broad range and RNA transfer. In *Cucurbita maxima* Phloem, Small RNA-Binding Protein 1 is capable of commodity. Only two coffee plant species,

- *C. arabica* and
- *C. canephora*, are economically important, accounting for 64 and 36 % of total global output, respectively.

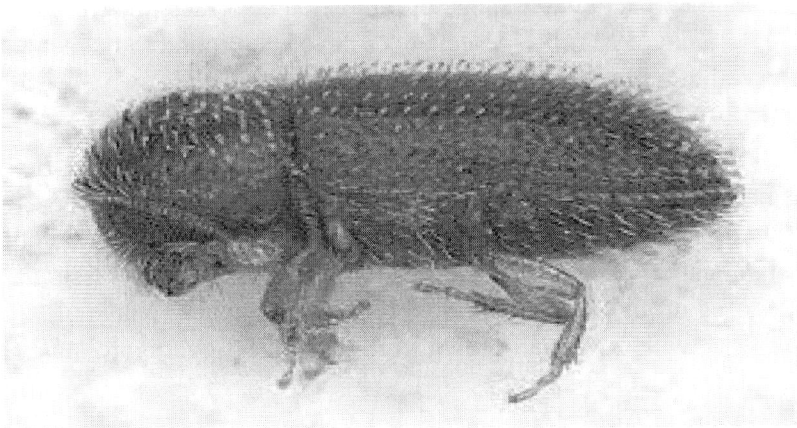

Figure 5.8. Coffee Berry Borer.

Source: Image by PestNet

The transcriptomes of *C. arabica* and *C. canephora* have been examined and characterised, allowing us, for the first time, to look at microRNAs in these two species. The first study on the discovery of microRNAs (miRNAs) in coffee (*C. arabica*) was recently published (Akter et al., 2014). This study used a well-developed, sophisticated, and comparative computational approach, as well as an EST-based homology search for the first time, to find miR393, a previously unknown large miRNA family with appropriate fold-back structures.

The newly found miR393 was anticipated to target genes encoding a transport inhibitor-like protein, transcription factors, DNA-binding proteins, and a GRR1-like protein. These target genes are engaged in a variety of biological processes, including responses to chitin, cold, salt stress, and hydration deprivation, among others. Several others conserved and non-conserved miRNAs, as well as their probable target genes, were identified and reported in a later study on C. arabica and C. canephora, which was published in Nature (Djami-Tchatchou et al., 2017).

5.7.4 Other Crops

Glycine max

Soybean is a globally prominent oil crop species. The ingredient has a high level of seed protein (40%) and a high level of oil (20%). While before 2008, the specific family of miRNAs and the specific group of 35 new miRNAs were both discovered, it was in 2008 that the specific family of miRNAs:

- miR156,
- miR159,
- miR160,
- miR164,
- miR169,
- miR172,
- miR390,
- miR393, and
- miR396

and the specific group of 35 new miRNAs including miR1511, miR1512, and miR482) were identified for the first time (Subramanian et al., 2008). This number now stands at 639 soybean miRNA sequences that have been submitted in miRBase 21.

Barley

Barley is one of the world's most significant crop species when it comes to overall productivity. Barley grain is consumed by humans and cattle (Curaba et al., 2012). It is an excellent model plant for the family Triticeae, especially in the realm of taxonomy, genomics, and breeding. At least 56 such as

- miR156/157,
- miR319,
- miR160,
- miR172,
- miR408, etc.

are discovered miRNAs in barley are orthologs in wheat or rice, which have been previously demonstrated to be expressed.

About 44 appear to be exclusively expressed in barley. The production of barley is comparable to other significant agricultural crops, such as rice and wheat, which have contributed to the surge in interest in miRNA research and the discovery of 69 miRNAs.

Potato

According to the UN's Food and Agriculture Organization, potato is the world's fourth most significant food crop, behind wheat, maize, and rice. The list of 48 potato miRNAs was established in 2009, and they were expected to target 186 different genes, including those for plant growth, metabolism, signaling, and stress resistance (Zhang et al., 2009). Through in silico methods, many conserved miRNAs were found in potato, and 567 miRNAs are available in the miRbase 21 (www.mirbase.org) repositories (Kim, 2011).

Orange

The fruit is best eaten fresh, and its juice has become popular as a drink because of its high vitamin C content. The peel is also used for skincare and in home cleaning goods, thus making it a great source of experimental material for studying the activities and regulatory mechanisms of miRNAs. *C. sinensis* miRNAs (csi-miRNAs) were found in leaves, stems, flowers, and fruits, and their anticipated gene targets were suggested by the researchers. The miRBase Team was able to identify over 120 orange miRNAs in the latest version of the database. miRNA expression regulation in orange root development can help stimulate root growth and help the plant tolerate both biotic and abiotic stressors, increasing the biomass and yield of the plant (Song et al., 2012)

Apple

As an economically significant fruit tree, apple cultivation is carried out globally, and its fruit is enjoyed by many people. Flavonoids and other phenolic chemicals are abundant in this fruit, which may be important in helping reduce the risk of chronic illness in humans.

miRNABase 21 discovered and described a total of four hundred and thirteen miRNAs (miRBase 21) from apple, all of which provide insight into how these molecules regulate processes. Previous research has shown that a lot of miRNAs are found in the phloem tissue and phloem sap (Djami-Tchatchou *et al.*, 2017)

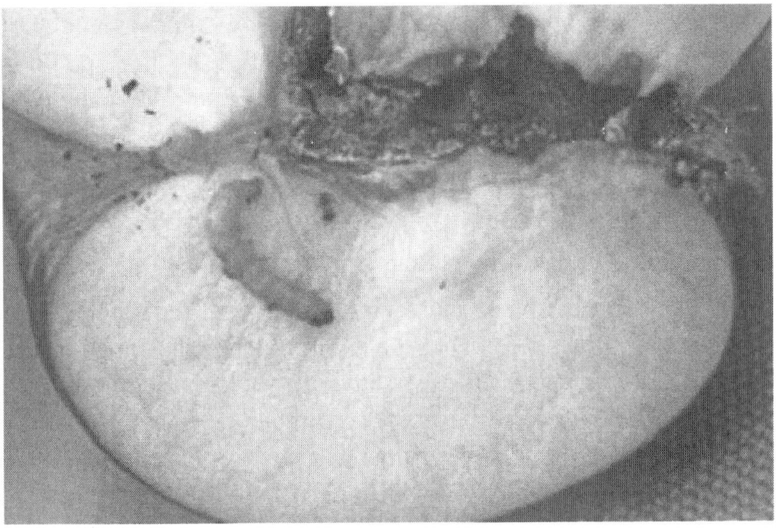

Figure 5.9. Codling moth larvae, *Cydia pomonella.*

Source: Image by Ward Upham, Kansas State University, Bugwood.org

Sugarcane

About 75% of the global sugar supply is derived from sugarcane. Renewable energy, such as ethanol, is increasingly significant in production. Biomass productivity and built-in confinement make it an attractive target for the development of sustainable biomaterial manufacturing. There is a very complicated polyploid genome of sugarcane in the world that has not yet been sequenced, and gene and genomic survey sequence databases help in looking for miRNAs.

A significant degree of similarity exists between a large percentage of sugarcane miRNA precursors and their putative sorghum orthologs. Additionally, because sugarcane and sorghum share a high degree of genetic similarity, using sorghum as a reference genome is advantageous for studies of sugarcane (Jannoo et al., 2007). The team of Zanca et al. (2010) found that the total number of targets discovered for 19 sugarcane miRNAs was 46, including many of the biological processes they govern.

5.9. CONCLUSION

The conclusion that can be drawn from the above chapter is that the plant volatiles or VOC's play a very crucial role by acting as chemical messages to help herbivore insects to locate host plant facilitating feeding, copulation, oviposition and helping parasitoids and predators to locate their prey species, serving as attractants for species-specific pollinators and even by inducing defense against pests by acting as feeding and/or oviposition deterrents.

REFERENCES

1. Akter, A., Islam, M. M., Mondal, S. I., Mahmud, Z., Jewel, N. A., Ferdous, S., et al. (2014). Computational identification of miRNA and targets from expressed sequence tags of coffee (Coffea arabica). Saudi J. Biol. Sci. 21, 3–12. doi: 10.1016/j.sjbs.2013.04.007
2. Ansari, M.S.; Moraiet, M.A.; Ahmad, S(2014) . Insecticides: Impact on the Environment and Human Health. In Environmental Deterioration and Human Health; Springer Netherlands: Dordrecht, The Netherlands; Volume 9789400778, pp. 99–123, ISBN 9789400778900.
3. Baek, D., Villén, J., Shin, C., Camargo, F. D., Gygi, S. P., and Bartel, D. P. (2008). The impact of microRNAs on protein output. Nature 455, 64–71. doi: 10.1038/nature07242
4. Buck, A.H.; Coakley et al., (2014). Exosomes secreted by nematode parasites transfer small RNAs to mammalian cells and modulate innate immunity. Nat. Commun. 5, 5488. [CrossRef]
5. Burmester, T. (2002).Origin and evolution of arthropod hemocyanins and related proteins. J. Comp. Physiol. B Biochem. Syst. Environ. Physiol. 172, 95–107.
6. Chow, et al., (2019)
7. Dutta, T.; Srivastava, S. S (2018) mall RNA-mediated regulation in bacteria: A growing palette of diverse mechanisms. Gene , 656, 60–72.
8. Djami-Tchatchou, A. T. *et al.* (2017) 'Functional roles of microRNAs in agronomically important plants-potential as targets for crop improvement and protection', *Frontiers in Plant Science*, 8(March). doi: 10.3389/fpls.2017.00378.
9. Fire, A.; Xu, et al. (1998(Potent and specific genetic interference by doublestranded RNA in *Caenorhabditis elegans*. Nature , 391, 806–811.
10. Gill, R.J.; Baldock,. et al. (2016) Protecting an Ecosystem Service. In Advances in Ecological Research; Academic Press: Cambridge, MA, USA,; pp. 135–206, ISBN 9780081009789.
11. Guleria, P., Mahajan, M., Bhardwaj, J., and Yadav, S. (2011). Plant Small RNAs: biogenesis, mode of action and their roles in abiotic stresses. Genom. Proteom. Bioinform. 9, 183–199. doi: 10.1016/S1672-0229(11)60022-3
12. Ham, B.-K.; Lucas, W.J.(2007) Phloem-Mobile RNAs as Systemic Signaling Agents. Annu. Rev. Plant Biol, 68, 173–195.

13. Huang, C.-Y.; Wang, H.; Hu, P.; Hamby, R.; Jin, H. (2019) Small RNAs—Big Players in Plant-Microbe Interactions. Cell Host Microbe, 26, 173–182.
14. Jactel, H.; Verheggen, F et al., (2019) . Alternatives to neonicotinoids. Environ. Int. **2019**, 129, 423–429.
15. Jannoo, N., Grivet, L., Chantret, N., Garsmeur, O., Glaszmann, J. C., Arruda, P., et al. (2007). Orthologous comparison in a gene-rich region among grasses reveals stability in the sugarcane polyploidy genome. Plant J. 50, 574–585. doi: 10.1111/j.1365-313X.2007.03082.x
16. Jiao, Y., Wang, Y., Xue, D., Wang, J., Yan, M., Liu, G., et al. (2010). Regulation of OsSPL14 by OsmiR156 defines ideal plant architecture in rice. Nat. Genet. 42, 541–544. doi: 10.1038/ng.591
17. Jones-Rhoades, M. W., Bartel, D. P., and Bartel, B. (2006). MicroRNAs and their regulatory roles in plants. Ann. Rev. Plant Biol. 57, 19–53. doi: 10.1146/annurev.arplant.57.032905.105218
18. Kim, H. J., Baek, K. H., Lee, B. W., Choi, D., and Hur, C. G. (2011). In silico identification and characterization of microRNAs and their putative target genes in Solanaceae plants. Genome 54, 91–98. doi: 10.1139/G10-104
19. L.; Jing, X. et al.(2012) Host Plant-Derived miRNAs Potentially Modulate the Development of a Cosmopolitan Insect Pest, *Plutella xylostella*. Biomolecules, 9, 602..
20. Llave, C., Xie, Z., Kasschau, K. D., and Carrington, J. C. (2002). Cleavage of scarecrow-like mRNA targets directed by a class of Arabidopsis miRNA. Science 297, 2053–2056. doi: 10.1126/science.1076311
21. Masood, M.; Everett, C.P.; Chan, S.Y.; Snow, J.W. et al., (2012) Negligible uptake and transfer of diet-derived pollen microRNAs in adult honey bees. RNA Biol.13, 109–118. [CrossRef] [PubMed]
22. Mello, C.C. Nobel Lecture. Available online: https://www.nobelprize.org/prizes/medicine/2006/mello/lecture/ (accessed on 23 February 2021).
23. Nazim Uddin, M.; Kim, J.Y. et al., (2013) Intercellular and systemic spread of RNA and RNAi in plants. Wiley Interdiscip. Rev. RNA, 4, 279–293.
24. Olina, A.V. et al., (2018) Proteins and Mechanisms of RNA Interference in Eukaryotes and Prokaryotes. Biochemistry , 83, 483–497.

25. Palatnik, J. F., Wollmann, H., Schommer, C., Schwab, R., Boisbouvier, J., Rodriguez, R., et al. (2007). Sequence and expression differences underlie functional specialization ofArabidopsis microRNAs miR159 and miR319. Dev. Cell 13, 115–125. doi: 10.1016/j.devcel.2007.04.012
26. Pantaleo, V., Szittya, G., Moxon, S., Miozzi, L., Moulton, V., Dalmay, T., et al. (2010). Identification of grapevine microRNAs and their targets using high-throughput sequencing and degradome analysis. Plant J. 62, 960–976. doi: 10.1111/j.0960-7412.2010.04208.x
27. Qin, Z., Li, C., Mao, L., and Wu, L. (2014). Novel insights from non-conserved microRNAs in plants. Front. Plant Sci. 5:586. doi: 10.3389/fpls.2014.00586
28. Santos, D. *et al.* (2021) 'Rnas on the go: Extracellular transfer in insects with promising prospects for pest management', *Plants*, 10(3), pp. 1–22. doi: 10.3390/plants10030484.
29. Schommer, C., Bresso, E. G., Spinelli, S. V., and Palatnik, J. F. (2012). "Role of microRNA miR319 in plant development. "MicroRNAs in Plant Development and Stress Responses"," in Signaling and Communication in Plants, Vol. 15, ed R. Sunkar (Berlin: Springer-Verlag), 29–47.
30. Song, C., Yu, M., Han, J., Wang, C., Liu, H., Zhang, Y., et al. (2012). Validation and characterization of *Citrus sinensis* microRNAs and their target genes. BMC Res.Notes 5:235. doi: 10.1186/1756-0500-5-235
31. Sun, G. (2012). MicroRNAs and their diverse functions in plants. Plant Mol. Biol. 80, 17–36. doi: 10.1007/s11103-011-9817-6
32. Thuringer, D.; Jego et al., (2016) Gap junction-mediated transfer of miR-145-5p from microvascular endothelial cells to colon cancer cells inhibits angiogenesis. Oncotarget, 7, 28160–28168.
33. Vogel, E.; Santos, D.; Mingels, L.; Verdonckt, T.-W.; Vanden Broeck, J.(2019) RNA Interference in Insects: Protecting Beneficials and Controlling Pests. Front. Physiol., 9, 1–21.
34. Wang, J. F., Zhou, H., Chen, Y. Q., Luo, Q. J., and Qu, L. H. (2004). Identification of 20 microRNAs from Oryza sativa. Nucleic Acids Res. 32, 1688–1695. doi: 10.1093/nar/gkh332
35. Wang, M.; Dean, R.A(2020) . Movement of small RNAs in and between plants and fungi. Mol. Plant Pathol.21, 589–601.
36. Watkins, D.; Arya, D.P. et al.,(2019)Regulatory roles of small RNAs in prokaryotes: Parallels and contrast with eukaryotic miRNA.

37. White, R.; Kumar, S. et al., (2020) Extracellular vesicles from *Heligmosomoides bakeri* and *Trichuris muris* contain distinct microRNA families and small RNAs that could underpin different functions in the host. Int. J. Parasitol., 50, 719–729.
38. Xia, K., Wang, R., Ou, X., Fang, Z., Tian, C., Duan, J., et al. (2012). OsTIR1 and OsAFB2 down regulation via OsmiR393 overexpression leads to more tillers, early flowering and less tolerance to salt and drought in rice. PLoS ONE 7:e30039. doi: 10.1371/journal.pone.0030039
39. Zanca, A. S., Vicentini, R., Ortiz-Morea, F. A., Bem, L. E. D., da Silva, M. J., Vincentz, M., et al. (2010). Identification and expression analysis of microRNAs and targets in the biofuel crop sugarcane. BMC Plant Biol. 10:260. doi: 10.1186/1471-2229-10-260
40. Zeng, J.; Gupta, V.K. et al., (2019) Cross-Kingdom Small RNAs Among Animals, Plants and Microbes. Cells, 8, 371.
41. Zhang et al., et al., (2012) Analysis of plant-derived miRNAs in animal small RNA datasets. BMC Genomics, 13, 381.
42. Zhang, S.; Hong, Z. et al., (2019) Mobile RNAs—the magical elf traveling between plant and the associated organisms. ExRNA **2019**, 1, 8.
43. Zhang, Y., Wang, W., Chen, J., Liu, J., Xia, M., and Shen, F. (2015). Identification of miRNAs and their targets in cotton inoculated with Verticillium dahliae by high-throughput sequencing and degradome analysis. Int. J. Mol. Sci. 16, 14749–14768. doi: 10.3390/ijms160714749
44. Zhang, Z., Yu, J., Li, D., Zhang, Z., Liu, F., Zhou, X., et al. (2010). PMRD: plant microRNA database. Nucleic Acids Res. 38, D806–D813. doi: 10.1093/nar/gkp818
45. Zhu, K.; Liu et al. (2017) Plant microRNAs in larval food regulate honeybee caste development. PLOS Genet.13, e1006946.

Chapter 6

Impact of Mitogen-Activated Protein Kinase in Plant Pest Interaction

CONTENTS

This chapter mainly focuses on the fact that a wide variety of environmental and developmental signals are converted into adaptive and pre-planned responses in eukaryotic cells via a complex pathway known as the MARK (mitogen-activated protein kinase) cascade. The downstream signals resulting from the cascades of MAPK kinases amplify signals by allowing the phosphorylation of three main groups of reversibly phosphorylated kinases, which then result in the phosphorylation of the substrate proteins to alter the activities of the proteins and influences a variety of responses, including genetic alterations. In addition to relaying and amplifying signals, MAPK cascades are engaged in signal transmission and amplification.

6.1. INTRODUCTION

Most herbivores have, for a long time, been insects, and plants have evolved and developed a wide range of methods to defend themselves against their invasion. While this is not true for all plants, for some species that are particularly vulnerable to insect herbivory, the precise pattern of tissue damage is shown to be associated with insect herbivory in select species.

Many plant species have developed a method for sensing specific insect oral secretions (OS) that are present in wounds, which allows them to begin a cascade of interrelated signaling pathways that ultimately culminates in the synthesis of a wide range of defensive compounds. Because of climate change and population growth, plants are forced to rescue themselves in order to meet their nutritional requirements.

Because a plant is an immobile species, it is unable to migrate in order to discover new sources of food, making it considerably more difficult to meet nutritional requirements. As a result of being able to detect stress more accurately, plants are more likely to respond properly and so survive.

Although several signaling cascades may have kinase components in common, the temporal and spatial limitations imposed on these pathways preserve their specificity, as well as additional methods, including cross-inhibition, feedback management, and scaffolding. A diverse range of plant processes is under the control of MAPK cascades, including stress responses, hormone release, and immunological responses, as well as developmental programs.

Through genetic studies, it has been discovered that several major MAPK components such as the *Arabidopsis thaliana* MAPKs MPK3, 4, and 6, as well as the MAP2Ks MKK1, 2, 4, and 5 are shared by different processes, including that of the Arabidopsis thaliana MAPKs MPK3, 4, and 6, as well as the MAP2Ks MKK1, 2, 4, and 5. MAPKs MPK3, 4, and 6, and MAP2Ks MKK1, 2, 4, and 5 have all were found in *Arabidopsis thaliana*.

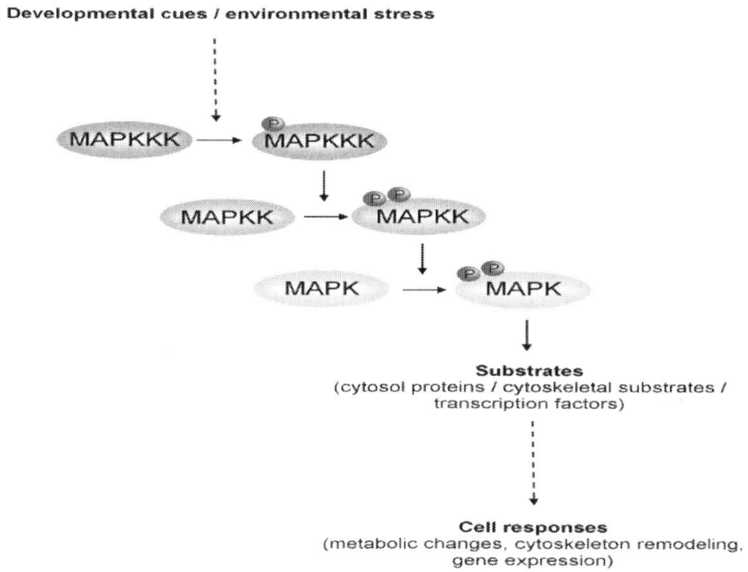

Figure 6.1. Schematic representation of the MAPK cascade.

Source: Image by Jagodzik et al., 2018

MAPK networks in plants regulate responses to four major stressors, including high salinity, drought, extreme heat, and infection, by functioning as signaling modules in eukaryotes. High salinity, drought, extreme heat, and infections are the primary stressors. A link has been shown between the activation of MAPK cascades and the response of plants to environmental stressors, in which they affect the expression of genes involved with plant hormones and influence the interaction with other environmental factors.

Current research on environmental stress responses is dependent on mitogen-activated protein kinase (MAPK). It is through MAPK phosphorylation that signaling proteins that are downstream of MAPKs are activated. There are several other signaling proteins that become activated in this process, including kinases, enzymes, cytoskeletal proteins, and transcription factors. In the MAPK cascade, three serine/threonine kinases are activated in succession:

- MAP kinase kinase kinase (MAPKKK),

regulated both neurogenesis and the synthesis of norepinephrine. Indeed, overexpression of WIPK led to increases in nicotine levels in tobacco plants when WIPK was over-expressed, can be silenced by RNAi, while silencing of WIPK resulted in reductions in nicotine levels in tobacco plants (as a result of WIPK expression being decreased) (Seo et al., 2007).

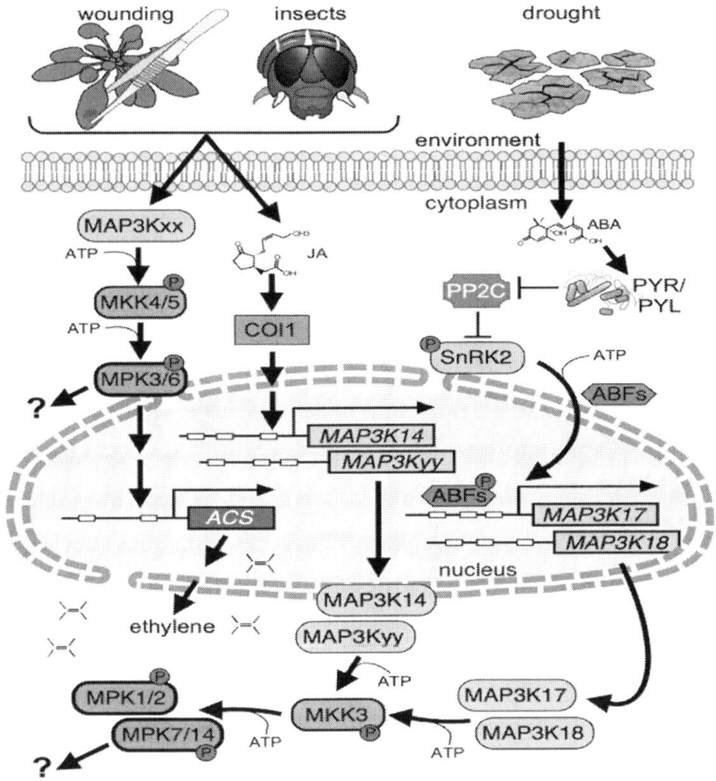

Figure 6.2. Activation of MAPK by Wounding.

Source:(Sözen et al., 2020)

Furthermore, through controlling MPK6 activity, it appears that JA homeostasis is maintained via a feedback loop. Clade A, sometimes known as stress-activated MAPKs, includes MAPK3 and MAPK6. Data also showed that members of the MAPK family whose research has yet to be completed might convey wound signals. The activation of the clade-D MAPK MPK8 occurs, as well as the activation of the clade-C MAPKs MPK1 and MPK2 (Takahashi et al., 2011).

6.3. MAPK CASCADES OF PLANT ORIGIN

The MAP kinase kinases (MAPKKs or MAP3Ks; also known as MAP2Ks or MEKs) are involved in the transduction and enhancement of input signals by MAPK cascades. There are three types of kinase involved in this process: mitogen-activated protein (MAP) kinase kinases (MAPKKs or MAP3Ks; also known as MAP2Ks or MEKs), MAP.

The MAPKKKKs discovered in plants have indeed been referred to as "MAP kinase kinases" in the same way that their cellular counterparts have been (Raja et al., 2017). First and foremost, activation of MAPKKKK or MAPKKK is accomplished by stimulation of plasma membrane receptors. A downstream MAPKK is subsequently activated as a result of MAPKK phosphorylating S/T-X5-S/T (where X is any amino acid).

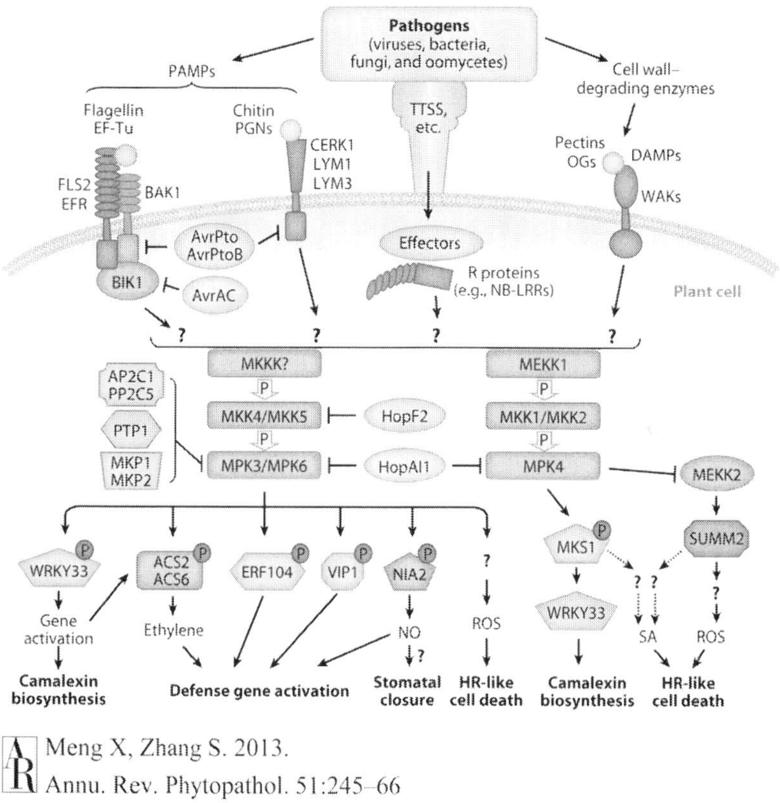

Figure 6.3. Working of MAPK cascades in signaling.

Source: Image by Xiangzong Meng,2013, Sözen et al., 2020

Because the MAPKK is a dual-specificity kinase, it can activate both types of receptors. Upon reaching the subsection VII and VIII activation loop in the catalytic domain, it phosphorylates the T-X-Y motif in the activation loop and between subdomains VII and VIII of the catalytic domain(Jagodzik *et al.*, 2018).

In the cytoplasm or nucleus, MAPKs are involved in the regulation of a range of other kinases, enzymes, cytoskeletal proteins, and transcription factors, as well as many other reactions. There are docking sites in the enzymes themselves, as well as docking sites in external scaffolding proteins, that allow for more efficient interactions between kinases.

A MAPK cascade is a sequence of phosphorylation events that occur one after another as a result of the action of the MAPK signaling pathway. It was discovered that approximately 110 genes in the *Arabidopsis thaliana* genome that encode members of the MAPK cascade kinase family, which is the largest number ever discovered. These genes are responsible for the production of:

- 20 members of the MAPK family,
- 10 members of the MAPKK family, and
- 80 members of the MAPKKK family, among other things (Raja et al., 2017).

The rice genome, for example, has genes for MAPK, MAPKK, and 75 MAPKKK, amongst other things (Wankhede et al., 2013). The MAPK cascade kinase genes are found in similar quantities in various plant species; for example, the tomato genome has:

- 17 MAPK,
- 5 MAPKK, and
- 89 MAPKKK genes, respectively (Wu et al., 2014).

As a result of these discoveries, it is possible that gene sets with similar sets of genes in other plant genomes, as well as the existence of an enormous range of MAPK cascade kinases throughout higher plant genomes, have been conserved throughout evolution (Jagodzik *et al.*, 2018).

Members of the MAPKK family include a large and diversified collection of MAPK cascade components. They include the following components. In Arabidopsis MAPKKKs, there are three primary categories that have been identified:

- MEKK-like (with 21 members),
- Raf-like (with 48 members), and

- ZIK-like (WNK [without lysine (K)] or WNK-like (with 11 members); alternatively, these MAPKKKs are referred to as WNK-like MAPKKKs (Danquah et al., 2014).

The MEKK-like kinase subfamily may be split into four subgroups, which are as follows: A1-A4 are the first four letters of the alphabet (MAPK Group, 2002).

A member of the protein kinase family, denoted as "AtMEKK1," "AtMEKK2," "AtMEKK3," and "AtMEKK4," is found in all Arabidopsis thaliana strains in the subgroup "A1," but no member of the kinase family has been identified in any of the other subgroups (MAPK Group, 2002). In addition, MEKK-like kinases from subgroup A1 have been identified in a variety of additional species.

6.3.1. MAP2Ks

MEKs and MKKs are classified into four groups (A-D), each of which has been studied in *A. thaliana* and rice (Hamel, 2006). Class A consists of the *A. thaliana* MKK1 as well as MKK2 proteins, which operate upstream of both the MAPK and MPK4 (Ichimura, et al., 1998).MKK2 has also been linked to reactions to cold and salt, and that MKK1 and MKK2 are involved in the regulation of innate immunity (Teige et al., 2004).

MKK6 was found to trigger MPK13, which was found in almost the same group (90). Particularly, we need to determine if the algal and early terrestrial plant homologs in group A interact and create cascades that respond to abiotic and biochemical stressors in conjunction with group B MAPKs in *A. thaliana*, as described further below.

Class B MKKs comprise *A. thaliana* MKK3, which has homologous proteins in rice, Selaginella and Chlamydomonas, as well as MKKs from other bacteria. The nuclear transfer factor (NTF) motif distinguishes kinases from other enzymes in this class (Hamel et al., 2006). It has been demonstrated that NTF domains act to facilitate the nuclear import of transport proteins, suggesting that plant MAP2Ks having NTF domains are engaged in cytoplasmic-nuclear mobility of proteins. Furthermore, MKK3 contributes to pathogen-induced circuits that are reliant on jasmonic acid (JA) signal transduction and are activated by the presence of pathogens. A sub-cellular level investigation of MKK3 and potential downstream MAPKs would be interesting in order to determine if they are colocalized, as well as to determine the processes of stimulation of JA-responsive MAPK modules.

It appears that groups C and D, which contain *A. thaliana* MKKs 4 and 5, as well as the remaining *A. thaliana* MKKs 7–10, are connected to a small group that contains 2 Physcomitrella and two Selaginella MKKs, respectively. All plant phyla appear to employ a smaller number of MKKs when contrasted to other MAPK subunits, and the survey discovered only a single MKK in each of the algae, Chlamydomonas and Volvox, which is consistent with previous findings.

This suggests that the very same MAP2K protein may be used by several MAPK modules at the same time. Genetic studies of *A. thaliana* have revealed that strongly linked sets of plant MAP2Ks possess activities that are quite similar. Examples include the proposal that MKK1 and MKK2 trigger the MAPK MPK4, whilst MKK4 and MKK5 operate upstream of MPK3 and MP6, seemingly in a redundant way, and are suggested to operate ahead of MPK3 and MP6 (Asai, 2002).

6.3.2. MAPKs

Like animal ERK kinases, plant MPKs feature phosphorylation motifs (TDY or TEY) in their activation loops (Fujita et al., 2006). Yet, we have not identified an annotated plant MAPK with such a TGY, or an annotated animal or yeast MAPK family member (like p38 and Hog1) that has a TPY (Group M, 2002).

A. thaliana MAK, which has a TxY domain and MAPK-related nucleotides in its catalytic domain (Moron et al., 1993) . Chlamydomonas crLF4 modulates flagellar size and phosphorylates the MAPK target MBP, whereas *A. thaliana* MAK modulates flagellar size and phosphorylates the MAPK precursor MBP *in vitro*. *C. elegans* has many homologs of CrLF4, one of which might be comparable to it owing to its effects on Cilil kinesin motors (Cristina, Petersen and Mundy, 2010).

CrLF4 has many homologs in *C. elegans*, one of which may be comparable to it due to its actions on cilil kinesin motors (Burghoorn et al., 2007). MAKs are involved in plant cell division and male germ cell development, and the pathways and activities of these MAKs parallel those of kinesins in flagellar, ciliar, and spindle cell functioning (Miyata et al., 1999). Plant MAK proteins feature a TxY motif typical of CDC2/CDK2 family kinases. However, it is necessary to show that MAK kinases constitute a distinct member of the MAPK family by functional investigations that insert MAK proteins into a MAPK module.

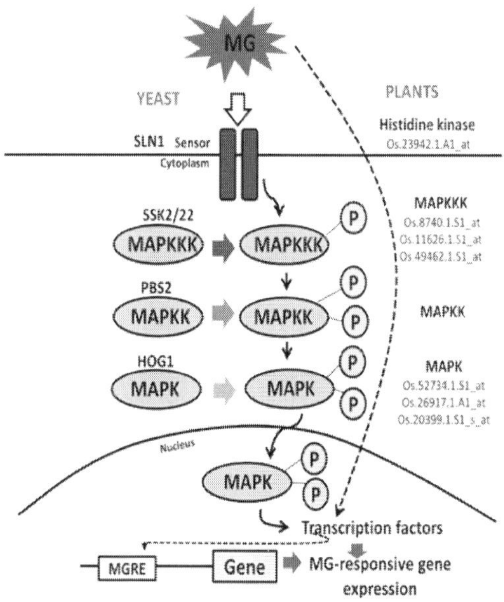

Figure 6.4. MAPK signaling cascade in yeast and plants.

Source: Image by Charanpreet Kaur,2015, Sözen et al., 2020

There are four *A. thaliana* MPK sub-groups (A–D; 53). Groups A, B, and C are known as TEY MAPKs, whereas Group D comprises TEY and TDY MAPKs. Included are MPK3 and MPK6 from the plant species Medicago sativa, rice, and poplar, as well as their orthologs which have been investigated in tobacco, *Medicago sativa*, and *O. sativa* (Cristina, Petersen and Mundy, 2010). Activated by different biotic and abiotic stressors, these MAPKs are involved in developmental processes. Early land plants, Selaginella and Physcomitrella, do not have obvious homologs in group A. MPK4, which is involved in pathogen defense and abiotic stress responses, is present in group B (Qiu, 2008).

MAP2Ks frequently include a C-terminal common docking domain that might be a docking target for MAPKs. Studies suggest that investigation of annotated MAPKs, it appears that ERK-related lineages gave rise to plant MAPKs, which were then passed on to early land plants in order to better adapt to changing conditions. Algae *C. reinhardtii* and *V. carteri* use eight MAPKs, but *Physcomitrella* and *Selaginella* utilize 10 MPKs.

A comparative study of the monocot and dicot MAPKs found that to prevent confusion over orthologs and paralogs in Arabidopsis, rice, and poplar, it would be necessary to use a new nomenclature to differentiate them. There are almost as many MAPK gene models found in poplar, relative to *A. thaliana*, as there are found in *O. sativa*, with the number of different splicing variants being greater in rice. Based on transcriptional research, they are believed to be the results of recent chromosomal duplications, which are observed to be linked to dramatic tissue and organ-level differences in expression. These findings support the notion of the fast development of specific plant MAPK functions (Hamel, 2006).

6.4. PEST ATTACK ON PLANT AND ACTIVATION OF MAPK CASCADE

Plants are effective in surviving biotic stress because they can quickly recognize, understand, and then deliver an appropriate response to the signal that has been presented to them. Stress caused by biological agents can have both short- and long-term repercussions. In addition to being understudied, events in the early seconds to minutes of life, which are responsible for the identification and activation of signal transduction pathways, are also understudied.

Damage to plant tissues caused by invading microorganisms and insect herbivory is inescapable, but the pace and severity of the injury can vary greatly depending on the organism. Herbivorous insects consume small leaf fragments on a continuous basis by cutting off and eating the tissues, which they do by cutting off and swallowing the tissues.

When these processes are initiated, they can be compared to a series of mechanical wounds, with saliva and bodily fluids being injected at the same time. When a single mechanical wound is applied, for example, the release of volatile compounds, there is little uniformity in the reaction. In Lima beans, continuous mechanical wounding results in about the same mix of volatiles as does herbivory, although the amounts generated are markedly different from one another. This indicates that while a herbivore-like wound is required, it is not sufficient in and of itself, and that both involved players must have a reciprocal chemical link in order to release the complete reaction chain (Maffei, Mithöfer and Boland, 2007).

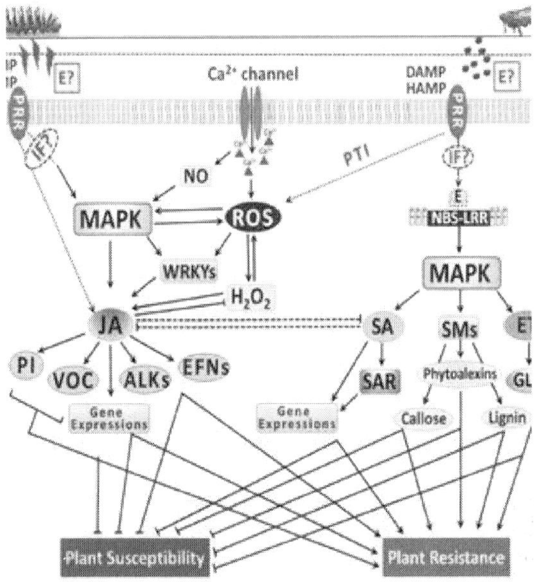

Figure 6.5. Activation of MAPK as Response to Pest Attack.

Source: Image by Satyabrata Nanda,2021, Sözen et al., 2020

As the sole component of the cell that has close contact with the surrounding environment, the plasma membrane serves as the sensory element, capable of recognizing changing conditions and initiating cascades of events that finally elicit responses.

Vm fluctuations or modification of ion fluxes at the cell membranes region are among the early cellular reaction to biotic and abiotic disturbances (Maffei et al., 2006). A rapid electrical signal termed as action potential is produced in response to attack pest triggered Vm alterations, and it travels throughout the plant body from the site where it was first detected to its destination (Maffei et al., 2006).

Vm alterations are often much slower than action potentials, which propagate up to 40 m sec1, which is too quick for the activity of a chemical signal with either phloem or xylem. The actual process usually takes approximately 5–6 min for such 6–7 cm in lima bean leaves, but it includes signaling molecules tr and trn (Peiffer et al., 2005).

The involvement of transcription factors, which can influence Vm both directly and indirectly through receptors, is particularly important in the early recognition between the host and disease or herbivore. A powerful depolarizing molecule such as hydrogen peroxide (H2O2), for example, can be triggered or transported by the eating insect.

The depolarization of the Vm is also caused by elicitors such as the peptide pep, as well as b-glucans and other oligosaccharides generated from the breakdown of pathogen cell walls (Shibuya, 2001). This one has been established that oral discharges of herbivores can induce ion fluxes and Vm depolarization.

The direct effect of channel activity on voltage (Vm) makes it plausible to infer that unknown elicitors govern the activities of different channels and, as a result, may be regarded essential variables in the generated electrical signaling. Ca2+-channels are the most important of them, as they are engaged in cell signaling the most (Maffei, Mithöfer and Boland, 2007).

6.4.1. Role of Ca2+

A second messenger in several plant signaling pathways, the calcium ion is well-known. The cytosol-free Ca2+ in healthy normal tissues is regulated at a range of around 100 and 200 nM. Compared to the apoplastic fluid, this concentration is approximately 104 times lower, and it is approximately 104 to 105 times lower in cellular organelles, providing the motivating factor for Ca2+ entry into the cytosol, where it functions as a stimulus.

Calcium signatures are used to refer to variations in [Ca2+]cyt that occur across time and space. These variations might manifest themselves as single Ca2+ transients, oscillations, or recurrent spikes that have a defined subcellular origin, lag time, magnitude, and frequency. Furthermore, various calcium signatures can be induced by the same signal based on the organ or the cell type within a tissue (Evans et al., 2001).

In response to transitory Ca2+ fluctuations, an array of Ca2+-binding proteins decodes the information contained in the changes, which results in a cascade of downstream consequences, involving associated protein phosphorylation and gene transcription patterns (Sanders et al., 2002).

There are two primary types of Ca2+-binding proteins, which are referred to be sensor relays and sensor responders, correspondingly. Calmodulin (CaM), CaM-related proteins, and calcineurin B-like proteins (CBL) are all examples of detector circuits that can work as a result of bimolecular associations. In Lima beans that had been successfully attacked with spider mites (*Tetranychus urticae*), as well as in leaves that had been exposed to the generated volatiles from injured plants, stimulation with a chelator of extracellular Ca2+ reduced the transcription of defensive genes (Arimura et al. 2000). Aequorin technology, which is based on bioluminescence, has been used to directly monitor calcium variations *in vivo* using fluorescent probes. A temporary rise in cytosolic Ca2+ in the area of the bite zone is observed when lepidopteran larvae feed on plants.

When plants meet herbivores, their [Ca2+]cyt increases primarily as a result of Ca2+ absorption from the extracellular media, Ca2+ mobilization from organelles, and/or a combination of the two. Mechanical injury, on the other hand, has no effect on the elevated [Ca2+]cyt due to insect salivary secretions.

In this context, it is still unclear if insect elicitors such as volicitin and inception are involved in Ca2+ influx in contrast to the relationship between Ca2+ and reactive oxygen species. Aside from the modulation of [Ca2+]cyt levels, nothing is documented well about the management of other potential ion channels by insect-derived elicitors, the substrate proteins that are impacted by pharmacological agents, or the potential for adverse that reflect changes in the Ca2+ concentration. Additional analysis of signal-sensitive ion channels, the connection of signal interpretation with ion channel activity, and the coupling of downstream components in hypothesized signal cascades are all obviously required (Maffei, et al. 2006).

6.4.2. Production of ROS

The production of reactive oxygen species (ROS) is a common response to pathogens. They are actively engaged in cell signaling and invasion, and they play a dual role: at low concentrations, they behave as second messengers and are part of the direct defense; at high concentrations, they are responsible for the entire direct defense. Besides the action of mechanical wounding, ROS is also produced in plant tissues during encounters with plant pests (Watanabe, 2001). It has been noted that pathogen-induced synthesis of ROS, specifically H2O2, occurs both locally and systemically in the leaves of several plant species (Orozco et al., 2001). Hormone levels rise in reaction to pest attacks as soon as the predators are still there. Moreover, the presence of H2O2 inside the plant as a result of pest attack could be effective in preventing secondary bacterial infections.

Orozco (2001) investigated that microbiological invasion of wound areas. Several ion channels have been shown to be affected by ROS action. H2O2 is released during the interaction between the herbivore *Spodoptera littoralis* and the legume Lima bean, which has been demonstrated to occur during herbivore feeding and, to a lesser extent, mechanical damage. The release of H2O2 results in a depolarization of the Vm canal, which involves Ca2+ networks. Lima bean leaves that have been mechanically damaged (Vm of approximately 140 mV) respond quickly to causative stimuli (Overmyer et al.,2003).

H2O2 is produced by trying to induce a significant Vm depolarization. Leaves that have been damaged by herbivores have a continuous decreased Vm (about 90 mV) and exhibit a significantly diminished or even non-responsive response to further H2O2 treatment. The lengthy depolarization of the Vm caused by H2O2 might be interpreted as an attempt by the insect to quiet the plant's system (Maffei, 2006). This is because the insect is continuously introducing a significant volume of H2O2 is injected into the injured tissue (Musser, 2006).

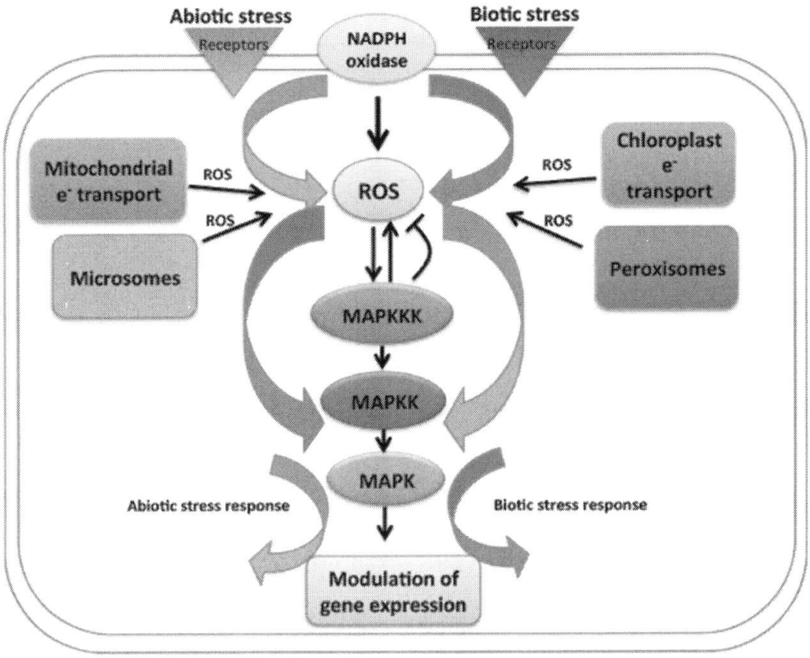

Figure 6.6. ROS Mediated MAPK.

Source: Image by Siddhi K. Jalmi,2015, Sözen et al., 2020

In order to deal with the increased amounts of H2O2, activity has been increased. For example, the presence of H2O2 recycling enzymes in plants, catalase production rises (Maffei, et al., 2006). Nitric oxide (NO), in addition to hydrogen peroxide (H2O2), is a transcription factor in plant response reactions. NO has the potential to act as downregulator of various wound-related signals and to promote wound healing (Delledonne, 2005). It lessens the formation of wound-induced hydrogen peroxide in the plant. Interactions between pathogens, jasmonic acid (JA) therapy resulted in a significant increase in NO generation, suggesting a self-amplifying mechanism, the JA–NO loop. It is also recognized that NO can cause a build-up of salicylic acid (SA), a key factor in the development of systemic resistance mechanisms the process of establishing resistance (Delledonne, 2005).

As a result, the function,of NO in responding to biotic stressors is being investigated. SA appears to have fine-tuned the process, which might cause JA to be inhibited. Biosynthesis and inhibit the initiation of the usual transcriptional machinery reaction by the defense. Even though the existence and activity of NO have not been determined, the JA-linked NO has been investigated in plant–insect interactions.

Given the circumstances, it is possible that production is also important in this situation. It is undeniable that jasmonates serve an important role. More information and specifics studies on the incorporation of NO into the signaling network are being conducted. It is important to have routes that are engaged in pest attacks (Maffei, Mithöfer and Boland, 2007).

6.4.3. Early Signal Transduction and Role of Kinases

MAPK cascades, which are key channels that control immune functions to both external and internal stimuli in organisms, are essential routes downstream of detectors and receptors. MAPKs appear to have a function in the signaling of biotic and abiotic stressors, pathogens, and secondary metabolites in plants, according to the available data. MARK (mitogen-activated protein Kinase) was extracted from the leaves of a variety of tobacco that encodes a mitogen-activated protein kinase (MAPK) whose mRNA began to concentrate in the leaves just minutes just after the infliction of a simple mechanical injury (Seo, et al., 1999).

Since suppressed plants have been unable to create JA and methyl-JA following injury, the protein kinase (WIPK) was found to be critical for JA synthesis and JA-induced reactions to be present in the plant. Increased buildup of SA and its glucoside was seen following wounding due to lack of WIPK activity at the injury site. Aside from that, it has also been shown that WIPK can induce transcription of the gene encoding an enzyme known as v-3 fatty acid desaturase (FAD7), which catalyzes the breakdown of linoleic acid to linolenic acid, which serves as a precursor to the amino acid JA (Kodama et al., 2000) On the other hand, it has been discovered that the WIPK and the SA-induced protein kinase (SIPK) share an upstream MAPKK, NtMEK2, and that they both associate with the calmodulin (CaM)-binding MAPK phosphatase (NtMKP1) in nicotine (Seo, et al., 1999). When the wound-induced protein kinase (WIPK) phosphorylates the NtWIF transcription factor in tobacco plants, the gene encoding is activated.

SIPKK (SA Induced Protein Kinase Kinase) activated NtMPK4, a tobacco homolog of *Arabidopsis thaliana* MPK4, in a study published in the journal Plant Physiology (Kodama et al., 2000). Transgenic tobacco plants have constitutively active NtWIF revealed a spontaneous buildup of expressions for the pathogenesis associated genes PR-1a and PR-2, as well as an increased degree of the antimicrobial saline (SA). Considering that NtWIF is situated downstream of WIPK, the findings imply that it is likely to have an impact on SA production (Waller, et al. 2006).

Although calcium-dependent protein kinases (CDPKs) are routinely implicated in signaling pathways in response to a variety of biotic and abiotic stressors, it is impossible to rule out the possibility that they may function as active protein cascades in herbivore and wound responses. Cellular CDPKs (also known as CDPKs) are a vast collection of serine/threonine kinases in plants as there are 34 isoforms in *Arabidopsis thaliana*.

Another connection with CDPKs (NtCDPK2), MAP kinases, and JA as well as ethylene was seen in potato (*Solanum tuberosum*). When the solanaceous generalist pest *Manduca sexta* attacks *Nicotiana attenuata* the *S. nigrum*, several kinases, CDPKs, and MAPKs have been shown to be activated, including many CDPKs and MAPKs (Schmidt, 2006).

6.5 MAPKS IN PLANT HORMONES SIGNALING

Tissue damage done by chewing pests, followed by the networks of plant growth regulators, represents a later part of the signalling pathways that play a role in signal transduction. A finely tuned reaction is managed to be achieved by using a combination of synergistic and antagonistic encounters, also known as cross-talk, that are dependent on both the herbivorous pest and the invaded plant.

This helps regulate gene expression and, as a result, regulates the manufacturing of protective metabolites in the attacked plant. To give an example, pest attack increases the synthesis of JA (10–15-fold), its precursor 12-oxophytodienoic acid and ethylene in many plants, while decreasing the levels of ABA and IAA (Schmelz, et al., 2003). It is possible to determine the temporal duration of the various phytohormones using GC-MS techniques following extraction and derivatization, which may be done after extraction and derivatization. The typical resting content of JA has been determined to be 10–40 ng g1 FW in various plants, and this volume rises 10–15-fold after the first 1–2 hours following injury in these plants. After injury to Lima beans, the first increase in JA was seen as early as 10 minutes after damage and reached a peak in the next 80–90 minutes. While trans-JA constituted 80 percent of total JA at rest, the physiologically more active cis-JA took over as the dominant form soon after injury, constituting an additional 80 percent of total JA. Mostly during feeding trial, a distinct spatial gradient of various oxylipins was seen to emerge (Baldwin, et al., 1997).

Immediately adjacent to the feeding site, the largest concentration occurred, and it extended 10–20 mm into the surrounding uninjured region. If regional gradients surrounding the biting zone are typical of JA and 12-OPDA, it needs to be seen if they are also typical of other species. Only a small rise in the amount of 12-OPDA, rather than an increase in the level of JA, was detected in the adjacent leaves.

The fact that 12-OPDA is not just a predictor of JA, but may also function as a signal that is not dependent on JA, suggests that its build-up may be crucial for the activation of a small subsection of defensive responses. Other oxylipins that have been produced at the same time as JA may possibly play a role in plant defense, although their function is less known at this time. A study of *N. attenuata* found that silencing the octadecanoid route led to plants that were more sensitive to pest invasion, showing the significance of this system for plant defense towards pests (Kessler, et al. (2004).

Microbiological infections are typically associated with significantly higher degrees of SA than those generated by insect feeding. The level of SA in corn (*Zea mays*) and Arabidopsis was only mildly elevated following the introduction of SA. Another immediate and active plant response to biotic stress is the synthesis of more ethylene.

In maize, ethylene synergizes with volatile emissions to increase resistance to pest herbivory and regulates a variety of additional defense mechanisms against pest herbivory. Finally, in Arabidopsis, the gene encoding gene AtMYC2 and the ethylene response factor 1 (ERF1) have been identified as critical components in the incorporation of stimuli originating from JA and ethylene through cross-talk, correspondingly (Garcia, et al. 2006).

6.5.1. Jasmonic Acid and Salicylic Acid Production by MAPKs

JA and SA are vital in plant growth, development, and defense (Hayat, 2007). SA is also crucial in plant growth, development, and defense. Additionally, SA is involved with SAR and reactions to abiotic stressors such as water, salt, and cold stress, in addition to its participation in wound healing. SA has been shown to affect the activity of Arabidopsis AtMPK3 and AtEDR1, as well as maize ZmMKK6-ZmMP4-1, in the laboratory. It's been demonstrated that ozone stress induces the expression of AtMPK3 in an SA-dependent way.

MPK3 appears to be a key crosstalk regulator of late immunological responses in the context of disease resistance (Mao et al., 2011). Aside from the previously described function of AtMPK3 in suppressing the transcription of defense genes both naturally occurring and triggered by flg22 (Galletti et al., 2011), AtMPK3 has been shown to be a specific inhibitor of flg22-induced SA deposition (Frei et al., 2014).

A major regulator of plant growth and development, the phytohormone JA also plays a significant role in the wounding response and stress-induced apoptosis (SAR) (Heil and colleagues, 2012; Hou et al., 2013; Wasternack and Hause, 2013). Although there has been documented crosstalk between JA and MAPK signaling, only a small number of research studies have described the nature of this relationship. Despite this, MAPKs have been shown to control JA production as well as the expression of genes that are reliant on As just an example, the tomato proteins SlMPK6-1 and SlMPK6-2 act as upstream regulators of JA production and transcription factors, but the tomato protein SlMPK6-1 does not do the same (Kandoth et al., 2007).

It has been demonstrated that silencing of SlMPK6-1 and SlMPK6-2 at the same time reduces JA production as well as the transcription of JA-dependent defense genes. JA, on the other hand, is a transcription factor that controls both MAPK function and MAPK transcription factor. It is found in Arabidopsis leaves, 1 hour after JA treatment that the activity of AtMPK1/2 kinase has been increased (Ortiz et al., 2007).

AtMPK9 and AtMPK12 are also implicated in JA-induced stomatal closure (Khokon et al., 2015; de Zelicourt et al., 2016; Lee et al., 2016), which is a process that occurs in response to JA. Reyna and Yang (2006) demonstrated that JA treatment could induce the expression of three different OsMPK genes in rice, though only at the transcript level (Reyna and Yang, 2006). Studies demonstrated that JA treatment could induce the expression of OsMPK7, OsMPK20-5 and the OsMPK16 genes in rice. JA treatment has been shown to increase transcript levels for BnaRaf30 in canola (Sun et al., 2014).

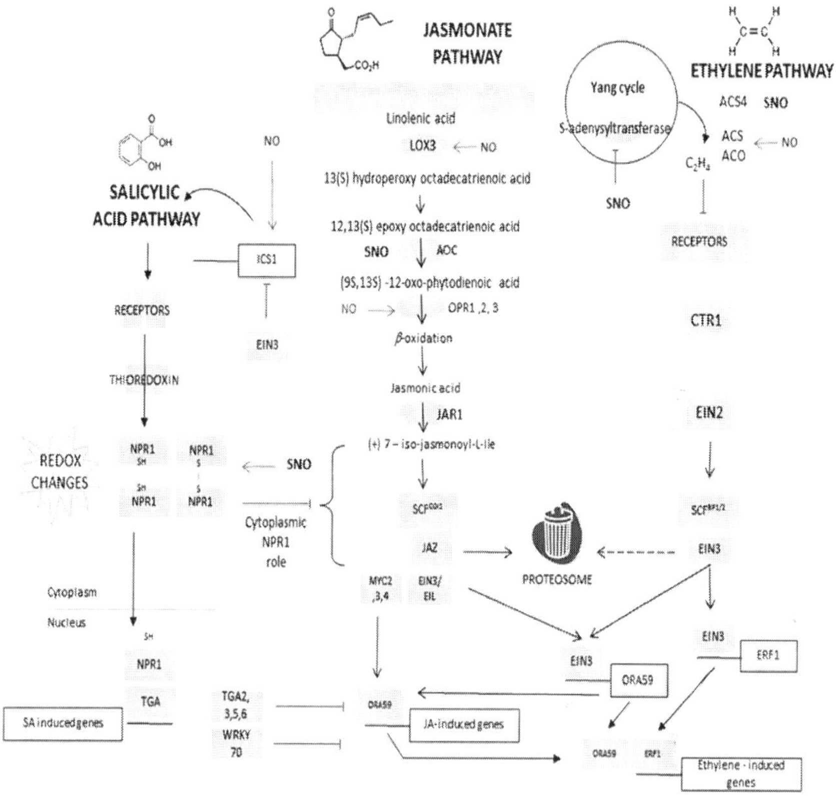

Figure 6.7. Plant Growth Hormone Signaling Pathway including JA, SA, Ethylene.

Salicylic acid (SA) is universally acknowledged for its function as a signal molecule in the immune system. Researchers discovered beyond a decade earlier that SA causes temporary stimulation of MAPKs in tobacco suspension cells, a discovery that led to the coining of the phrase SA-induced protein kinase. Researchers discovered beyond a decade earlier that SA causes temporary stimulation of MAPKs in tobacco suspension cells, a discovery that led to the coining of the phrase SA-induced protein kinase.

A. thaliana roots revealed that phosphorylation of MAPKs is inversely proportional to the number of SA applied. *A. thaliana* MPK6 is the homolog of SIPK. The increased expression of WIPK leads in higher JA synthesis, despite the fact that conflicting evidence suggests that stimulation of WIPK and SIPK, either by highly conserved tobacco MEK2 or diterpenoid therapy, may not result in higher JA manufacturing (Cristina, Petersen and Mundy, 2010).

Other MAPK cascades, in contrast to WIPK and SIPK, may be involved in controlling JA or SA production and responses, according to one interpretation of this phenomenon: MPK6, but still not MPK4, is a transcription factor that is selectively stimulated by JA. Also involved in JA signaling by microarray research was the MKK3 protein, and transgenic organisms that produce an inducible or constantly active MKK3 indicated that MPK6 was specifically activated in consequence to JA.

Another research, on the other hand, found that this hypothesis could not be verified in plants, which was disappointing. It was suggested instead that MKK3 might connect with and activate MAPKs in the evolutionary Group, which comprises MPKs 1, 2, 14, and most importantly MPK7 regarding H2O2, as demonstrated by this work. Additionally, ABA and hydrogen peroxide (H2O2) stimulate the activation of MPK1 and MPK2, and their expression is indeed a late JA and injury response. Furthermore, the fact that these MAPKs are co-localized in vascular tissues suggests that they function particularly in the vasculature. However, further research is required to identify the specific functions of such kinases inside vascular cells, despite the fact that this may relate them to systemic reactions (Cristina, Petersen and Mundy, 2010).

Wang et al., (2010) discovered that in maize, the MAPK ZmMPK3-2 (ZmMPK3) responds to different signaling chemicals, including JAs and SA. Also apparent in tomato is that the SlMKK4 (SlMKK2) and SlMKK9 (SlMKK4) genes appear to be engaged in both the jasmonate (JA) and salicylate (SA) signaling pathways (Li., et al., 2014).

Defending plants against diseases relies on signals from SA and JA. These numerous examples of MAPK cascade kinases demonstrating either the importance of collaboration in plants respond to injury or a joint role in signaling for JA, SA, or both show how critical it is for plants to collaborate in this manner. More research is needed to learn more about the intricacies of this partnership and how it boosts plant resistance to infections.

6.6. CONCLUSION

The conclusion that can be drawn from the above chapter is that MAPKs prioritize certain downstream substrates once they are active, including other kinases, enzymes, or transcription factors. Research has examined whether other protein kinases may activate MAPKs as well and whether this is the case. While prior research has been overwhelmingly focused on the functions of the MPK3-MPK4-MPK6 kinases and the other kinases and targets with which they interact.

28. Mao, G. *et al.* (2011). Phosphorylation of a WRKY transcription factor by two pathogen-responsive MAPKs drives phytoalexin biosynthesis in Arabidopsis. Plant Cell 23, 1639–1653. doi: 10.1105/tpc.111.084996
29. MAPK Group (2002). Mitogen-activated protein kinase cascades in plants, a new nomenclature. Trends Plant Sci. 7, 301–308. doi: 10.1016/S1360-1385(02) 02302-6
30. Mitho¨fer, A. et al. (2005) Effects of feeding *Spodoptera littoralis* on lima bean leaves. II. Continuous mechanical wounding resembling insect feeding is sufficient to elicit herbivory-related volatile emission. Plant Physiol. 137, 1160–1168
31. Miyata,Y.. *et al.* (1999). Molecular cloning and characterization of a novel member of the MAP kinase superfamily. *Genes Cells* 4:299–309
32. Moran, W.*et al* (1993). Molecular cloning of two novel protein kinase genes from *Arabidopsis thaliana. Biochim. Biophys. Acta* 1216:9–14
33. Musser, R.O. et al. (2006) Ablation of caterpillar labial salivary glands: technique for determining the role of saliva in insect–plant interactions. J. Chem. Ecol. 32, 981–992
34. Orozco-Cardenas, M.L. et al. (2001) Hydrogen peroxide acts as a second messenger for the induction of defense genes in tomato plants in response to wounding, systemin, and methyl jasmonate. Plant Cell 13, 179–191
35. Overmyer, K. et al. (2003) Reactive oxygen species and hormonal control of cell death. Trends Plant Sci. 8, 335–342
36. Peiffer, M. *et al.*(2005) The host plant as a factor in the synthesis and secretion of salivary glucose oxidase in larval *Helicoverpa zea*. Arch. Insec. Biochem. 58, 106–113 Physiol. 51, 754–766. doi: 10.1093/pcp/pcq037
37. Qiu, et al. (2008). *Arabidopsis* MAP kinase 4 regulates gene expression through transcription factor release in the nucleus. *EMBO J.* 27:2214–21
38. Raja, V.*et al.*(2017). Abiotic stress: interplay between ROS, hormones and MAPKs. Environ. Exp. Bot. 137, 142–157. doi: 10.1016/j.envexpbot.2017.02.010
39. Reymond, P., Bodenhausen, N., Van Poecke, R.M.P., Krishnamurthy,V., Dicke, M., and Farmer, E.E. (2004). A conserved transcript pattern in response to a specialist and a generalist herbivore. Plant Cell 16:3132–3147.

40. Roux PP, Blenis J. 2004. ERK and p38 MAPK-activated protein kinases: a family ofprotein kinases with diverse biological functions. Microbiol. Mol. Biol. Rev. 68:320–44
41. Sanders, D. et al. (2002) Calcium at the crossroads of signaling. Plant Cell 14, S401–S417
42. Savatin, D. et al. (2014) Wounding in the plant tissue: The defense of a dangerous passage.
43. Schmelz, E.A. et al. (2003) Simultaneous analysis of phytohormones, phytotoxins, and volatile organic compounds in plants. Proc. Natl. Acad. Sci. U. S. A. 100, 10552–10557
44. Schmidt, S. *et al.*(2006) Systemin in Solanum nigrum. The tomato-homologs polypeptide does not mediate direct defense responses. Plant Physiol. 142, 1751–1758
45. Seo, S. et al. (1999) Jasmonate-based wound signal transduction requires activation of WIPK, a tobacco mitogen-activated protein kinase. Plant Cell 11, 289–298
46. Seo, S. et al., (1995). Tobacco MAP kinase: A possible mediator in wound
47. Seo, S. et al., (1999). Jasmonate-based wound signal transduction requires activation of WIPK, a tobacco mitogen activated protein kinase. Plant Cell 11: 289–298.
48. Shibuya, N. and Minami, E. (2001) Oligosaccharide signalling for defence responses in plant. Physiol. Mol. Plant Pathol. 59, 223–233 signal transduction pathways. Science 270: 1988–1992.
49. Stafstrom JP, Altschuler M, Anderson DH. (1993). Molecular cloning and expression of a MAP kinase homologue from pea. Plant Mol. Biol. 22:83–90
50. Takahashi, F. *et al.* (2011). Calmodulin-dependent activation of MAP kinase for ROS homeostasis in Arabidopsis. Mol. Cell 41: 649–660.
51. Teige, M. et al. 2004. The MKK2 pathway mediates cold and salt stress signaling in *Arabidopsis*. *Mol. Cell* 15:141–52
52. Waller, F. et al. (2006) Expression of a WIPK-activated transcription factor results in increase of endogenous salicylic acid and pathogen resistance in tobacco plants. Plant Cell Physiol. 47, 1169–1174
53. Wang, X. et al. (2010). Two coupled components of the mitogen-activated protein kinase cascade MdMPK1 and MdMKK1 from apple function in ABA signal transduction. Plant Cell

54. Wankhede, D.*et al.* (2013). Rice mitogen activated protein kinase and mitogen activated protein kinase interaction network revealed by in-silico docking and yeast two-hybrid approaches. PLoS One 8:e65011. doi: 10.1371/journal.pone.0065011
55. Watanabe, T. et al. (2001) Wound-induced expression of a gene for 1-aminocyclopropane-1-carboxylate synthase and ethylene production are regulated by both reactive oxygen species and jasmonic acid in *Cucurbita maxima*. Plant Physiol. Biochem. 39, 121–127

Chapter 7

Insecticidal Proteins and Their Importance in Controlling Plant Pests

CONTENTS

This chapter of the book explains the role of insecticidal proteins in plants and their significance in controlling plant pests. This chapter also provides highlights on the concept of insecticidal protein expression, as well as the destruction caused by the insect pests to the plants. This chapter also addresses the role of insecticidal proteins that are present in plants. This chapter further sheds light on the significance of BT cotton such as a history of BT cotton along with the working of BT proteins in plants.

Then the chapter revolves around the various advances that have been made in BT toxins insecticidal capacity. This chapter explains the role of plant lectins, mode of action of plant lectins, and various safety assessments necessary for the lectins.

7.1. INTRODUCTION

Agricultural production suffers a $17.7 billion loss per year as a result of the negative effects of insect and pest predation (Oliveira et al., 2014). There are many kinds of insects such as Lepidoptera

- Rice yellow stem borer,
- Chickpea Pod borer,
- Cotton boll worm
- Pigeon pea,
- Rice weevil,
- Multiple pulse grains bruchids
- Cotton boll worm,
- Pigeon pea,
- Cabbage Maggot,
- Cutworms,
- Chickpea aphid,
- Pigeon pea
- Mustard aphid
- Colorado Potato Beetle
- Mexican Bean Beetle
- Chickpea aphid,
- Green leafhop aphid (Paul and Das, 2021)

As a result, such predators are susceptible to chemical insecticides, which make them expensive and environmentally hazardous. As a result, there is a growing demand for naturally occurring biopesticides that are less expensive and less environmentally destructive. In addition to a diverse range of microbial pesticides, many other natural environmental assets can be found.

Utilizing reliable biomolecules in crops will become increasingly simple in the future, as new advancements in genetic engineering technology become available. This will make it easier to protect crop plants against pest damage. In order to use biopesticides effectively, a better understanding of the biology, feeding behavior, and evolutionary history of the pest, as well as the specific actions of the biopesticides, is required. This allows for the precise determination of when and where the chemical compound should be administered in order to increase the effectiveness of the molecule in question.

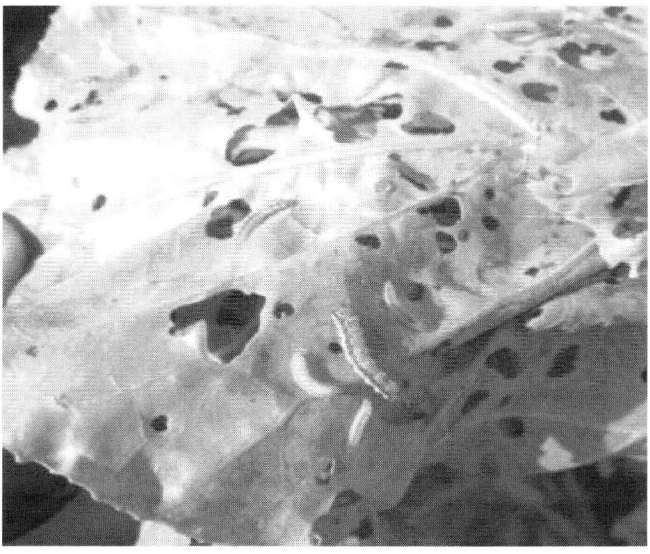

Figure-7.1. *Evergestis rimosalis larvae.*

Source: (Cabbage, Broccoli & Other Cole Crop Insect Pests | Home & Garden Information Center (clemson.edu))

It is estimated that Bt strains, together with their insecticidal crystalline proteins (ICPs)/Bt proteins, account for 90–95 percent of the world's largest bio-insecticides, with these substances having been in use for more than 50 years. In China, a total of 276 Bt microbial formulations are registered as of 2007 (Huang et al., 2007) Sep insect toxic substances and *Serratia entomophila* are two bacteria that produce biologically produced insecticides, which are also known as Sep and Bin insect toxic substances, respectively. In addition to producing powerful insecticidal toxins, the bacteria *Xenorhabdus* and *Photorhabdus* spp., which are members of the Enterobacteriaceae family, have been shown to increase the number of entomopathogenic nematodes.

Because of the efficacy, safety for humans and non-target organisms, and favorable natural consistency performance of Bt microbial preparations, genetically modified crops are genetically modified. As a result of increased farm income, it significantly reduced pesticide use, and significantly reduced emissions of greenhouse gases (GHGs) (Paul and Das, 2021).

Insecticide-resistant crops received the greatest benefit, as their pesticide application costs were reduced as a result of the reduction in pesticide application costs. It was found that insecticidal active ingredients used on insect-resistant maize and cotton decreased by 48 percent and 26 percent, respectively, in both developed and developing countries, according to the report.

Through the use of no-till methods, it is possible to reduce greenhouse gas emissions. No-till farming, on the other hand, has nothing to do with Bt crops. It is anticipated that lower pesticide applications will result as a result of the use of Bt crops, which will improve fuel efficiency. When input costs fall and crop yields rise, farm profits rise as a result of the reduction in costs of production and the increase in agricultural production.

In 2012, revenue from insect-resistant maize increased by $6.7 billion, while revenue from insect-resistant cotton increased by $5.5 billion. More than 75% of the additional revenue generated by cotton farming is generated in developing countries. Since that time, additional insecticidal elements have been discovered and reported, including plant lectins, protease inhibitors, and amylase inhibitors.

Phlopterids are a class of insects that cause only minor damage to crop species, in contrast to homopterans, which cause significant damage to a wide range of plant species by feeding on the phloem sap of the host plant species and transmitting disease-causing viruses from one plant species to another while feeding.

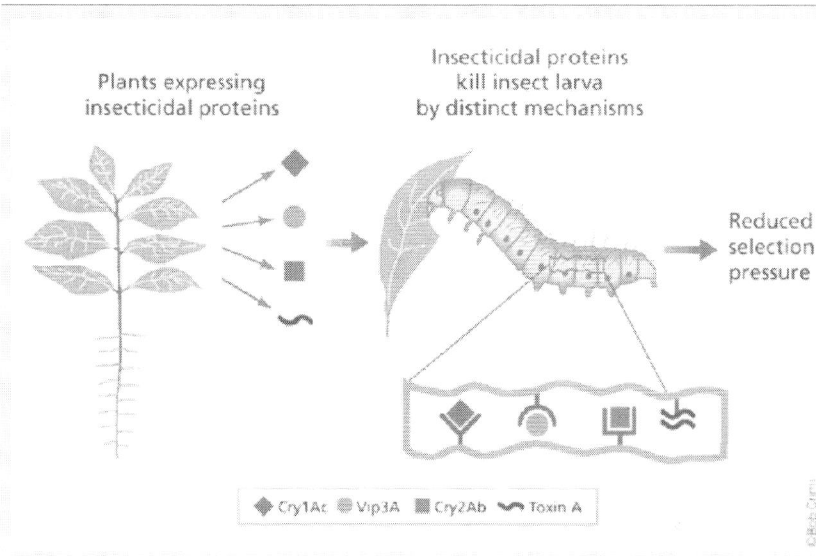

Figure 7.2. Insecticidal Proteins Expression.

Image Source: (Muhammad Wajid Javed, 2018)

These bugs do not come into contact with Bt toxins. Due to the common eating behavior of sucking up the free sugars and amino acids from the plant sap, they do not rely on the protease or amylase types of digestive enzymes as heavily as other animals. There is, however, a successful insect-eating homopteran group of sucking insects that have discovered an efficient plant lectin protein that recognizes and binds to carbohydrate moieties, which they have named the homopteran group.

Target insects have glycosylated receptors for lectins that are capable of binding to BBMVs on their midgut brush border membrane, which is found on the brush border membrane of their midgut (Bandyopadhyay, et al., 2001). Newer bioinsecticides and genetically modified crops for pest management must be successfully introduced into the market, which means that regulatory, economic, and social issues must all be taken into consideration (Roy et al., 2002).

Many crops suffer from the presence of the cotton bollworm (*Helicoverpa armigera*) and the armyworm (*Spodoptera litura*), which are among the most ferocious lepidopteran pests on the planet. There are several significant pest species in the Helicoverpa genus of the Noctuidae, notably *H. armigera, H. zea,* and *H. assulta*, which are all members of the Noctuidae family

(Behere, G. et al., 2007). *H armigera* is capable of causing crop potential losses estimated to be in the billions of dollars each year (Smith et al., 2014).

In addition to attacking over 180 species from 45 different families, including fiber and food crops, this pest has a broad host range. The destruction caused by *H. armigera* has been decreased to almost non-existent levels with the introduction of Bt crops. This pest has acquired resistance to Bt crops containing the Cry1Ac toxin, due to the capacity of insects to swiftly adapt and learn new management techniques.

The establishment of resistance in *H. armigera* against Cry1Ac was verified by *in vitro* investigations in 2003, and following that, several examples of resistance against Cry1Ac were reported (Tabashnik et al. 2003). *Spodoptera litura*, another member of the Noctuidae family, is likewise a global insect herbivore with a wide range of prey to which it may adapt.

A unique migratory capacity allows it to increase its population distribution around the world (Rauf *et al.*, 2019). Since it rapidly acquired resistance to chemical control techniques, *S. litura* has been extremely hard to control and manage. As an alternate strategy for managing *S. litura*, Bt toxins such as

- Cry2, Cry1FA,
- Vip3A, and
- Vip3b have all been produced. It has been demonstrated that these toxins have the capability to be employed as first-line management techniques in transgenic plants (Rauf *et al.*, 2019).

7.2. DESTRUCTION CAUSED BY INSECT PEST

Plants suffer significant losses as a result of insect pests, and it is anticipated that far more than 10,000 bug species cause major harm to essential food crops (Dhaliwal et al., 2007)

When it comes to species that live on earth, insects are by far the most varied group. Insect pests account for less than 0.5 percent of all recorded species of insects, and they inflict damage to plants both directly or indirectly during their quest to get food.

Almost all of a plant's parts, including the roots, stems, bark sprouts, leaves, buds, flowers, and fruits, are utterly destroyed by a variety of bug species that invade the plant in a variety of ways, each with its own method of infection. Natural disasters destroy approximately one-fifth of the world's agricultural production. More than 10,000 kinds of insects are recorded as being present each year. As a result, farmers and government organizations must first analyze the production losses caused by different types of insects before developing management methods to combat these losses. It is difficult to provide quantitative statistics on yield losses caused by insect pests since they vary depending on the cultivar, the density of the pest population, the timing of pest attack with respect to the crop phenology, and the traditional practices used.

A thorough investigation by the study discovered that, given the large use of synthetic chemical pesticides, losses due to insect pests rose in the post-green revolution era back to the non-green revolution era, when considered from a worldwide perspective.

Figure 7.3. Garden Slug.

Source: Common Insects and Associated Pests Attacking Bedding Plants and Perennials | Ohioline (osu.edu)

In India, a well-known homopteran pest, *Lipaphis erysimi* commonly known as mustard aphid, causes considerable yield loss in the *Brassica juncea* (Indian mustard) plant, this is one of the major sources of food oil in the world, accounting for 12 percent of total edible oil production worldwide.

Based on the level of the infection and the phase of the plants where the aphid attacks, this pest can inflict a 10–90 percent production loss on mustard in India (Rana et al., 2005). Furthermore, harmful viruses that aphids transmit while eating cause yield loss to increase. It is currently a common practice to use far higher amounts of toxics, such as insecticides, against insect pests than was formerly the case. Non-target organisms (as evidenced by the extinction of predators, parasites, pollinators, the environment, and human health) and target creatures (as evidenced by pesticide resistance and pest comeback) have all suffered as a result. For this reason, researchers have designed certain crop products to generate insecticidal proteins.

For example, as in the case of *Bacillus thuringiensis*, Bt, which has a focused effect on lepidopteran insects and has no adverse effects on other species in the environment (Sanahuja et al., 2011). It is possible that Bt crops will continue to be threatened as a result of the adaptation of insects, and that many lepidopteran pests may evolve resistance to Bt crops.

Figure 7.4. Japanese beetle.

Image Source: Common Insects and Associated Pests Attacking Bedding Plants and Perennials | Ohioline (osu.edu)

7.3. INSECTICIDAL PROTEIN

It has been discovered that there are a significant number of proteins that have insecticidal potential. A number of them, such as Cry proteins and plant lectins, have demonstrated promising effects on a wide range of insect pests, as well as the ability to create transgenic plants possessing insecticidal characteristics(Tripathi and Mishra, 2016).

Bt plants generate vegetative insecticidal proteins (Vip), which are the second generation of insecticides, during the vegetative development stage of their life cycle (Estruch et al., 1996). Vip1, Vip2, and Vip3 are all members of this protein family. It is worth noting that VIP1 and VIP2 are binary toxins that are effective for segments of the coleopteran group, whereas the Vip3 toxins are especially effective for lepidopterans (Estruch et al., 1996).

There are roughly 82 different types of vegetative insecticidal protein genes that have been discovered and cloned to this point in time. According to the amino acid sequence similarity of the encoded amino acids, these genes may be divided into three groups, eight subgroups, 25 classes, and 82 subclasses, respectively.

The Vip proteins, which exhibit no resemblance to endotoxins (Lee et al., 2003), have emerged as a significant special generation of insecticidal proteins with substantial implications for the control of insect populations. Vip causes intestinal paralysis in insects as well as total lysis of the gut epithelium. It was investigated that that Vip3A causes deadly toxicity in the larvae of *Spodoptera frugiperda*, and that Vip3Aa14 causes fatal poisoning in the *Spodoptera litura* larvae (Bhalla et al., 2005).

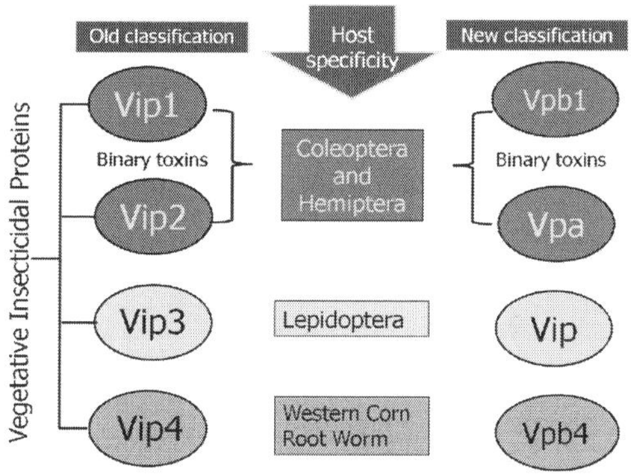

Figure 7.5. VIP family protein.

Image Source: (Kaur, 2021)

7.4. BT COTTON

7.4.1 History of Bt Cotton

In addition to soil, *Bacillus thuringiensis* (Bt) can be found in a wide range of other habitats, including deserts and prairies. It was discovered in 1901 by the Japanese researcher Ishiwata Shigetane, who was investigating the causes of an illness that affected silkworms at the time.

Afterwards, in 1911, the German scientist Ernst Berliner isolated Bt from flour moth caterpillars that were gathered in Thuringia, Germany, and re-isolated it in the laboratory (hence the species name). After further investigation, Berliner discovered that the Bt bacteria were only poisonous to particular insect larvae and not to others.

The first attempt to utilize Bt as a pest control tool was not made until 1928, and even then, it was unsuccessful (University of California San Diego Aroian Lab). During this first event, the bacteria was utilized to ward off the European corn borer (*Ostrinia nubilalis*), which has historically been such a frequent and extremely harmful maize pest in the United States. This resulted in the development of Sporine, the first commercial Bt-based biopesticide, which was reintroduced in France in 1938 \and became widely available the following year.

Throughout the years, Bt-based biopesticides have proven to be an effective pest control technique, and they are now a prevalent pest control strategy in sustainable production as well (Lambert, 1992). By the 1990s, hundreds or even thousands of strains had indeed been discovered, each of which was harmful to a diverse spectrum of insect and pest species.

Nonetheless, when the very first genetically modified maize produced with genes from Bt became available in 1995, it marked a watershed moment in agricultural history. Plants containing Bt genes have gained ground since then, and now account for 81 percent of all maize and 84 percent of total cotton acres grown in the United States respectively (Roh, 2007).

7.4.2. Working of Bt Protein

It has been discovered that the soil bacterium *Bacillus thuringiensisis* during spore formation generates protein insecticidal and that these endotoxins have insecticidal properties (Schnepf et al., 1998). Cry and cyt are belonging to multiple genetic families that are members of this group of endotoxins. According to research, cry proteins are poisonous to insects belonging to the orders of the Lepidoptera, Coleoptera, Hymenoptera, and Diptera (De et al., 2001).

Cyt toxins, on the other hand, are usually located in Bt strains that are effective towards Diptera, with a few outliers of Cyt proteins that are active against Coleopteran larvae described. There are now 74 Cry classes and 3 Cyt groups with far more than 700 individual participants, all of which have been characterized and categorized (Crickmore et al., 2014).

The toxicity of Bt toxins to a variety of insects, as well as the ease with which the gene could be isolated from the bacterial plasmid, made the production of the very first insect-resistant transgenic plants possible. Every Bt strain plasmid produces a toxin with a particular mechanism of action against insects, which is unique to that strain.

There are about more than 150 distinct Cry toxins that have all been isolated, cloned and evaluated for toxicity against a variety of destructive insect species. During the spore formation stage of Bt, endotoxins produce and deposited as protoxins in crystalline form, which are then excreted.

Following solubilization, such protoxins are released in the insect midgut where they are split off at the C-terminal portion, resulting in the release of a 66-kDa active N-terminal poisonous peptide with a harmful effect.

Cry1Ac transgenics have made significant strides forward in recent years, with the design of synthetic versions with codon reconfiguration to eliminate the putative polyadenylation sequences and introduce codon utilization that is favorable for high expression in dicotyledonous plants being one of the most significant developments in the field (Perlak et al., 1990). So far, the majority of genetically engineered crops that have demonstrated insecticidal action have consisted mostly of the toxin Cry1Ac in GM cotton and Cry1Ab in GM maize, respectively.

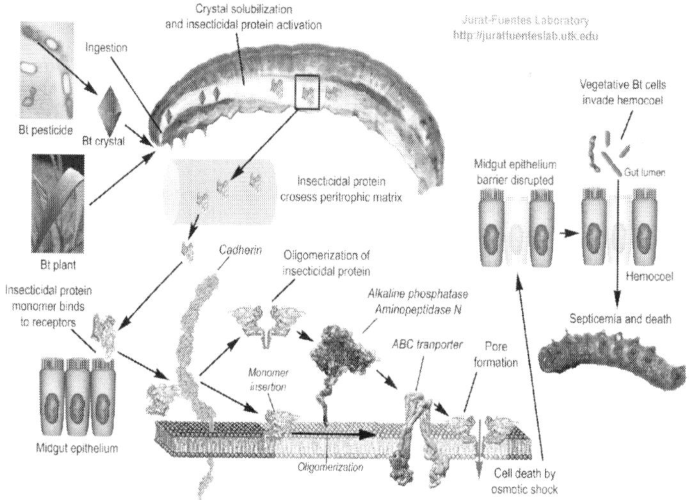

Figure 7.6. Working of Bt Cotton.

Image Source: Dr. Juan Luis Jurat-Fuentes (Department of Entomology and Plant Pathology) (utk.edu)

The effectiveness of a specific toxin, on the other hand, is typically restricted to a small number of closely related species. Some insect populations (for example, the diamondback moth) have developed resistance to the Bt toxin after being exposed to it for an extended period of time.

The development of resistance in a certain insect pest population indicates that Bt will no longer be effective in reducing that pest population. It is widely acknowledged that the extensive use of Bt in genetically modified plants would increase the selection pressure on insect populations, resulting in faster development of insect resistance to Bt. Some other bacteria, such as those produced from the bacterium *Bacillus cereus* and the bacterium *Photorhabdus luminescens*, also serve as a source of insect-specific toxin genes for research (Tripathi and Mishra, 2016).

On the basis of their structure and function, the Cry proteins can be divided into three main groups: the 3d-Cry toxins, the Mtx-like Cry toxins, and the Bin-like Cry toxins. The complex 3d-Cry toxins constitute the biggest category, and they contain the bulk of Cry virulence factors that have been utilized to alter plants in order to confer insect resistance.

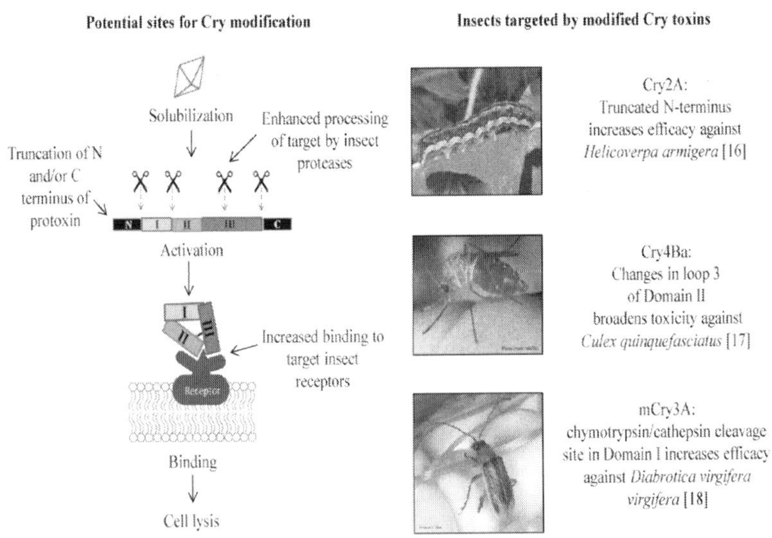

Figure 7.7. Cry Protein Modifications.

Image Source: (Benjamin R. Deist, 2014)

More than 40 distinct varieties of cry toxin (from Cry1 to Cry55) are classified under the 3d- cry toxin category, with several subgroups within each type (Cry1Aa, Cry1Ab ,etc.). The 3d-Cry toxins are structurally globular molecules that have three different regions or domains that are linked by a single linker to form a single molecule. The first domain of the Cry protein is important for pore creation, while the other two domains are crucial for the toxin's ability to attach to receptors and its insect specificity.

The Cry protoxin is ingested by susceptible larvae and is solubilized under alkaline conditions (pH > 9.5) in the insect midgut, where it is stimulated by chymotrypsin-like proteases, yielding a toxic segment of nearly 60 kDa, leading to a three-domain configuration (Johnston KA, L.et al. 1995). A very complicated process ensues when the activated form comes into contact with the particular receptors found on the midgut cells.

A contact with aminopeptidase N (APN) and alkaline phosphatase (ALP), which are both very frequently occurring proteins that are attached to the membranes with the help of glycosylphosphatidylinositol attachment point in a low-affinity association, is seen between the monomeric 3D cry toxin and these proteins.

Furthermore, certain *B. thuringiensis* and *B. cereus* strains generate a third class of insecticidal proteins known as Vip which are produced by the bacteria B. *thuringiensis* and *B. cereus*. Vip proteins are identical to Cry proteins in the way that they kill insects by attaching to their midguts, but the mode of action is distinct. Vip1, Vip2, and Vip3 are examples of this class of proteins. Vip causes intestinal paralysis in insects as well as total lysis of the gut epithelium.

Vip proteins are produced by the bacteria during the plant growth phase of its life cycle then released into the surrounding medium. When compared to other Vips, the well-designed Vip3A produces deadly toxicity in the larvae of *S. frugiperda*, while Vip3Aa14 induces lethal toxicity in the larvae of *P. xylostella*. In the VIP 1A gene, there is a 100 kDa protein encoded, which is synthesized from its N-terminal end. When this procedure is completed, an 80 kDa product is produced, and it has been demonstrated that this product is poisonous to western corn root warp larvae when this process is completed (Paul and Das, 2021).

7.5. ADVANCES IN BT TOXINS INSECTICIDAL CAPACITY

The discovery of the Bt toxin gene has resulted in the development of novel poisons that have the potential to be utilized for insect control and resistance control. Several key pests, unfortunately, exhibit limited vulnerability to the known Cry toxins, and the emergence of resistance to Cry toxins in many insect pests is jeopardizing the continued use of Bt technology in many situations. Another option is to genetically modify the insecticidal action of a Cry toxin to increase its potency.

Moreover, several genetic engineering techniques to enhance Cry toxicity have been examined and discussed (Bravo et al. 2013). The following section will discuss recent advances in the engineering of Cry or Cyt toxic substances that have led to increased toxicity, modification of toxin specificity, or the development of methods to overcome Cry toxin resistance.

The specificity of Cry toxins is primarily determined by their ability to recognize proteins found in the larval intestine (Bravo et al. 2011). The development of toxins to bind novel receptor molecules, as well as the increase in binding affinity for existing specific receptors, will give poisons with new pest specificities or greater toxicity. The mapping of binding areas in toxins and sensors is indeed critical information that will aid in the development of the toxins and receptors in order to increase their pesticide effectiveness throughout time.

In addition to the identification of larval gut proteins, domains II and III of 3d-Cry toxins are also involved in the recognition of other poisons. An evolutionary study of the 3d-Cry family indicated that domain III swapping among distinct Cry toxins had a role throughout the natural development of these proteins over the course of their evolutionary history (de Maagd et al. 2001).

There are domain III transactions with many distinct Cry toxins, which have resulted in an increase in the toxicity of the Cry toxins. They have demonstrated that this is a viable strategy for *in vitro* evolution of Cry toxins in a variety of situations (Bravo et al. 2013). The toxicity of *S. cerevisiae* was increased by 6-fold when a hybrid protein including domains I and II from Cry1Ab and domain III from Cry1C was constructed, as opposed to Cry1C, exigua (de Maagd et al. 2000).

In the instance of domain II, it's been demonstrated that accessible loop portions of this domain are important in receptor binding, as previously stated (Bravo et al. 2011). As a result, the use of loop exchange between different toxins is likely to be an effective approach. In order to reduce harm (Bravo et al. 2013).This strategy is illustrated by a good example of the alteration of the mosquitocidal selectivity of Cry4Aa caused by the substitution of the II loop 3 domain sequence with the loop3 sequence of Cry4Ba (Abdullah et al. 2003). *Culex sp.* became toxic to the hybrid protein as a result of its toxicity and *Aedes aegypti*, which has retained its toxicity (Abdullah et al. 2003).

In addition, it has been demonstrated that mutations in domain II loop sequences might result in the production of poisons with enhanced toxicity. Mutagenesis of loop sections of insecticidal proteins to alter insect selectivity was used in the development of the Cyt2Aa toxin, which is yet another example.

Cyt2Aa was modified by introducing a peptide sequence that binds to an aphid gut APN into the various exposed loop regions of the protein. Some of the hybrid proteins were found to be toxic to the pea aphid *Acyrthosiphon pisum* as well as the green aphid *Myzus persicae* after being tested on them (Chougule et al. 2013).

Then there are a few examples of high-throughput systems for the development of Cry poisons to conclude this section has been explained in detail. Phage display enables the presentation of many variations at the same time.

In the study of the toxins on the surface of the phage, as well as the selection of binders to brush border membranes or purified receptors, we were able to identify additional Cry toxins with increased toxicity against diverse insect species, which we have used to develop new insecticides (Bravo et al. 2013).

The Cry1Ac toxin was recently evolved to bind a non-cognate cadherin-like receptor using a new phage-assisted continuous evolution method (PACE), which rapidly generates high-affinity protein-protein interactions. With the help of this system, it is possible to improve the toxicity of the toxin while simultaneously overcoming resistance to Bt toxins (Bardan et al. 2016).

Finally, in order to counter Cry1Ab and Cry1Ac were designed to be resistant to Cry1A toxins, whereas Cry1A was created to be susceptible. After cadherin binding, the amino-terminal end of the protein is cleaved. Soberón et al (2007) demonstrated that these modified toxins (Cry1AbMod or Cry1AcMod) form oligomers in solution and kill Cry1A-resistant insects that are related to distinct resistance mechanisms.

All in all, identifying and mapping the rate-limiting steps of Cry toxicity in different insect pests, in addition to identifying and mapping the binding regions involved in receptor recognition, will likely result in the development of strategies to modify Cry toxins in order to change specificity, increase toxicity, enhance and induce counter resistance (Fiuza, Polanczyk and Crickmore, 2017).

7.6. PLANT LECTINS

Lectins are a non-immune protein family with at least one free domain that may bind to particular carbohydrates, whether basic monosaccharides or more extensive glycans (Peumans et al., 1995). Such protein molecules have been identified in plants, viruses, bacteria, invertebrates, and vertebrates, especially mammals, and are extensively dispersed throughout nature. Plant lectins are present in high quantities in specific tissues such as seeds, bark, and bulbs. Plant seeds from the Leguminoseae family are high in lectins (Paul and Das, 2021).

More than 230 three-dimensional configurations of lectins have indeed been uploaded in the 3D Lectin databank with 40 percent of these structures derived from legume seeds. In spite of the variations in carbohydrate-binding specificities between typical lectins from diverse plant sources, there is a high degree of structural similarity among typical lectins (Rouge´ P, et al., 1987). When supplied with fake food or expressed in transgenic plants, a variety of plant lectins have been found to impair the performance of insects belonging to several orders, including the Lepidoptera, Coleoptera, and Hemiptera (Vandenborre et al., 2010). It is believed that lectins bind with various glycoproteins or glycan structures in insects and that this interaction causes interference with a variety of physiological processes in these animals (Vandenborre et al., 2011).

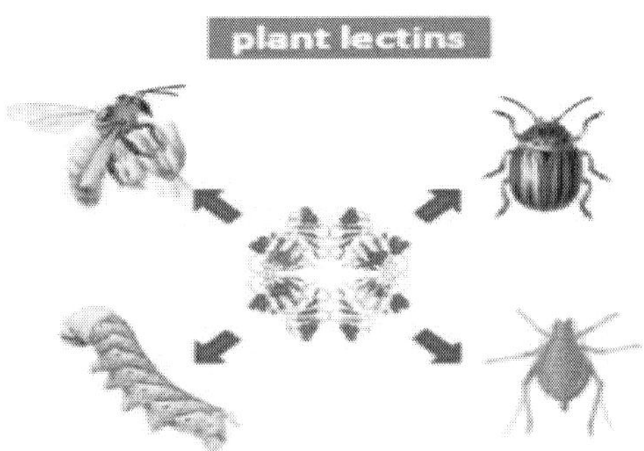

Figure 7.8. Plant Lectins.

Source: (Giaani Vandenborre, 2011)

Lectins include at least one carbohydrate-binding domain with varying sugar specificities, and diverse glycan structures are present in the bodies of insects, resulting in a large number of possible victims for lectin binding (Vandenborre G et al., 2010).

As a result, it is impossible to anticipate the precise mechanism of action of each lectin, and it is much more difficult to comprehend the diversity in insect toxicity as a result of exposure to diverse plant lectins in varied environments. Most lectins, regardless of their specificity, exhibit similar effects on informal processing like fertility, growth, and maturation.

In recent years, researchers have concentrated their efforts on developing more effective methods of controlling sap-sucking homopteran pests that are resistant to Cry toxins derived from the bacteria *Bacillus thuringiensis*. Since they are able to establish a large population in a short period of time, these insects have a significant detrimental influence on commercially important agricultural plants (Paul and Das, 2021).

Some mannose-binding lectins from plant species have risen to prominence as possible insecticides, particularly against sap-sucking homopterans, and have garnered significant attention. In addition to having insecticidal action against the Coleoptera and Lepidoptera groups of insects, plant lectins, especially mannose-binding monocot lectins, are extremely poisonous to the Homoptera group of insects, according to research. This requires the binding of the non-catalytic domains of the plant lectins to particular glycoprotein receptors on the insect gut epithelium in a reversible manner before it can be effective.

Plant lectins have an effect on a variety of physiological processes in insects, including fecundity, growth, and development. There has also been evidence of disruption in iron metabolism and amino acid metabolism, as well as changed eating behavior.

They are less sensitive to Bt or other spray formulations because they penetrate the leaves rather than eating on the leaf surface, but they are more vulnerable to plant lectins because they feed on the lectins of the plant (Paul and Das, 2021).

7.7. MODE OF ACTION OF PLANT LECTINS

Plant lectins serve a variety of biological functions, but their anti-insect activity has piqued the interest of researchers, as it is one of the well-studied insecticidal proteins (Lam et al., 2011). Growing interest in the utilization of plant lectins in pest control methods has increased in recent years, thanks to the emergence of plants that are resistant to insect assault.

In transgenic plants, the application of lectins has had beneficial results, particularly for crops that have been engineered to express Cry toxins of *B. thuringiensis (Bt)*, which have demonstrated resistance to sap-sucking insects. Furthermore, it has been demonstrated that the consumption of lectins in cultures grown or the production of lectins in transgenic plants reduces the activity of insects related to a variety of orders, including the Lepidoptera, Coleoptera, and Hemiptera, among several others (Vandenborre et al., 2011).

Despite extensive research on lectin toxicity, the precise mechanism by which lectins function in providing insect resistance remains unknown. Their entomotoxic effect is thought to be mediated mostly through interactions with various glycoproteins and glycan structures found in insects, which may affect a variety of physiological activities in these animals.

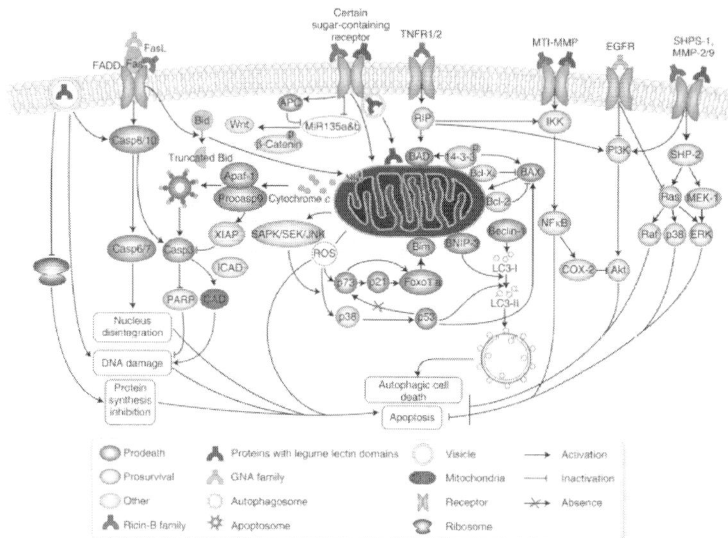

Figure 7.9. Mode of Action of Plant Lectins.

Source: (Ana Cristina Ribeiro, 2018)

In light of the fact that lectins contain at least one carbohydrate-binding domain and differing sugar specificities, as well as the wide diversity of glycan structures found in the bodies of insects, the number of potential targets for lectin binding is enormous.

As a result, it is impossible to anticipate the precise mechanism of work of each lectin, and it is much more difficult to comprehend the diversity of insect toxicity as a result of exposure to diverse plant lectins in varied environments (Macedo, Oliveira and Oliveira, 2015).

In general, antinutritional effects are frequently seen as a result of lectin intake, and these effects have the potential to impact a variety of biological factors in insects, including larval size, fertility, pupation, and survivability. A prolongation of the larval development cycle and adult emergence is also frequently seen.

Observe that one of the primary features of an insecticidal protein is its resilience to the proteolytic breakdown in the insect's stomach, which is a trait that should not be overlooked. Plant lectins have a high degree of resistance to the digestive enzymes of insects.

Examples include the fact that *Moringa oleifera* lecitin (cMoL) is resistant to degradation by *Anagasta kuehniella* protease enzymes for as long as 12 hours and *Olneya tesota* lectin (PF2) is resistant to digestion by *Z. subfasciatus* digestive enzymes lasting up to 24 hours. After incubation with midgut extracts of *S. littoralis* for up to three days, NICTABA demonstrates even greater resistance and is not digested (Felton, et al., 2005).

For a long time, researchers have concentrated their efforts on developing more effective methods of controlling sap-sucking hemipteran attacks on a variety of key crops. These insects have a significant detrimental influence on plants, which is exacerbated by their extraordinarily quick population expansion.

The effects of plant lectins on Hemipteran bugs have yielded some intriguing discoveries in terms of the method of action of these compounds. In recent years, the snowdrop lectin (also known as GNA or *Galanthus nivalis* agglutinin) has gotten a lot of interest because of its deadly effects on hemipterans, which are a type of insect pest that is significant for the agricultural industry (Bhatia, V. et al., 2011).

GNA binding to carbohydrate moieties on the surface of cells in the midgut epithelium of brown rice planthoppers was examined (*Nilaparvata lugens*). Furthermore, GNA induced morphological alterations in the epithelium, including the destruction of the microvilli brush boundary area, which was previously present.

Also present in the fat bodies, the ovarioles, and the hemolymph are its widespread systemic effects. GNA has been detected in all three of these locations. Also, when insects were fed on transgenic rice plants that expressed GNA, a honeydew deficit was found in *N. lugens* (2.0 percent of total soluble protein). Due to this, insects experience a drop in their survival and total fertility, as well as developmental delays.

ASAL, a mannose-binding lectin that has also been manufactured for a wide range of transgenic species of plants, such as tobacco and *B. juncea*, is another toxic lectin in the *Allium* genus (Dutta et al., 2005). Garlic lectins, ASAI and ASAII, formed as recombinant proteins, have an insecticidal impact if eaten by pea aphids (Fitches et al., 2008).

The survival of the larvae was negatively affected by a dose-dependent effect of these lectins, with death occurring most rapidly at the maximum lectin concentration (2.0 mg/mL). Two membrane-associated aphid gut proteins were identified as lectin binding receptor proteins, alanyl aminopeptidase N and sucrase. This means that the microvillar membrane of the insect's midgut is associated with 15% of the total gut protein in *A. pisum*. Additionally, it also combines significantly with GNA and ConA.

The insect's aminopeptidase interaction, which is linked to amino acid intake, is not hindered by the lectin, but even so, it inhibits aphid progression. Both these insect digestive enzyme activities, such as α-amylase, have also been shown to be affected by plant lectins, including α- and β-glucosidases and trypsin-like enzymes.

The coupling of GNA and ASAL to crucial midgut receptors, like ferritin and NADH quinone oxidoreductase, has also been discovered in other investigations. The two mannose-specific N-glycolyltransferases *C. esculenta* has a unique tuber agglutinin (CEA) that has been found to bind to various vacuolar ATP synthase receptor proteins identified as Bemisia tabaci sarcoplasmic reticulum type calcium ATPase and ATP synthase β subunit, along with heat shock protein 70, ATP synthase β subunit, and clathrin host-encoded protein from *Lipaphis erysimi*.

With these lectins being bound, there are likely to be changed in the metabolic system, likely to result in stunted growth, infertility, and death(Macedo, Oliveira and Oliveira, 2015).

It has been shown that plant lectins have an anti-insect action against lepidopteran pests. About 16 lectins isolated from 15 plant families were shown to negatively affect the legume pod-borer, *Maruca vitrata*, in a screening study. At least one of the parameters influenced by lectin ingestion was larval longevity, size, eating capacity, pupation, survival rate, and reproduction. The GNA lectin was among the most potent, and it had a noticeable influence on all tested parameters. A great deal of research has already confirmed the deleterious effects of GNA on various lepidopterans.

The larvae of the *L. oleracea* insect were fed an unnatural food intake and on transgenic potatoes that produced GNA, and their growth duration were both significantly reduced. *Eorreuma loftini*, the sugarcane cutworm, shows diminished larval viability, survival rate, and female reproduction when GNA is expressed in transgenic sugarcane.

While the harmful effects of GNA were not observed on *Diatraea saccharalis*, the deleterious effects of GNA were also not noted on *D. saccharalis*. Thus, the results also demonstrated that larvae that were fed GNA grew better throughout the first generation, but this was not found in the second generation (Macedo, Oliveira and Oliveira, 2015).

Despite the abundance of studies showing GNA can cause adverse effects, these side effects may be dependent on the amount of lectin and the type of insect. A number of investigations into the insecticidal properties of the *Bauhinia monandra* leaf lectin have been done, all with regards to the cowpea weevil (*Callosobruchus maculatus*), the Mexican bean weevil (*Zabota subfasciatus*), and the Mediterranean flour moth (*A.kuehniella*).

Bruchid survival in small doses (0.3% – 0.5% w/w) differed from *A. kuehniella*, where no effects on survival were seen in larvae fed 0.3% – 0.5% w/w. The decrease in *A. kuehniella* larval weight was 40% at this concentration. Ferritin was found to be a receptor for GNA inside the larval midgut of the predatory shrimp *S. littoralis*, as described previously in *N. lugens*. GNA, the authors theorized, may affect insects' iron metabolism because of interactions with proteins found in the digestive tract, such as ferritin, which impedes the animals' absorption of iron. HHA, a lectin found in Amaryllis hybrid bulbs, was shown to bind to the brush border of midgut cells in another study conducted by Caccia et al (2012). Labeled lectin is found to bind to the cell membrane of columnar cells, as well as being internalized.

HHA crossing the gut epithelial boundary and entering the insect hemolymph is implied by this information. They showed that FITC-HHA internalization is accomplished through clathrin-assisted endocytosis. Lectin-containing diets may reflect differences in gene expression in insects. *Drosophila melanogaster* larvae that consumed a WGA-rich diet upregulated 61 transcripts, including many that bind chitin.

They reasoned that because WGA caused structural damage to the PM, the midgut made more chitin in order to fix or replace the damaged

PM. Both these unique gene clusters have been known to be correlated with digestive enzymes and energy metabolism. The comprehension of the complex manner in which plant lectins and pests interact might be advanced by analyzing gene expression.

Alternatively, since the peritrophic membrane (PM) is mainly made up of chitin-microfibrils and glycoproteins, it may be a possible lectin target site in the gut region. They do not have a PM, while the hemipterans do not. A diet containing lectins causes PM deformities and microvilli disruption in lepidopterans.

Ostrinia nubilalis, a European corn borer, was fed on a diet containing wheat germ agglutinin (WGA) and grew with hypersecretion of unorganized PM (A) and disintegration of microvilli (M). WGA and its interaction may cause PM voids, which may cause food particles to pass via the PM and penetrate the microvillar brush border (Macedo, Oliveira and Oliveira, 2015).

There were no differences noted in the features of PM formation and structure for *M. sexta*, and thus there was no indication that these characteristics affect how WGA interacts with chitin as well as other modules of the PM. WGA causes obstruction in the microvilli of *Drosophila melanogaster*, which can cause abnormalities in the receptor structures. A further example is *Dioscorea batatas* agglutinin (DB1), which is classified as a mannose-binding lectin. It binds strongly to the larvae of *Helicoverpa armigera*.

A study by Walski et al. (2014) shows those three structurally related lectins that have different molecular sizes have diverse toxic effects. SNA-I and SNA-II, with a molecular weight of 240, 32, and 15.5 kDa, including both, were studied. SNA-II was the most toxic to the *Tribolium castaneum* cell line, followed by SNA-I, and lastly, RSA. In other words, RSA was most toxic when the lectins were mixed into the larvae diet at 2% (w/w). In order to discover how lectins, differ in their toxicity, the researchers discovered that RSA transited the PM more readily than SNA-II.

Additionally, both lectins were discovered to be immune to proteolysis, whereas SNA-I was discovered to be broken down in the gastrointestinal tract. While the monomeric form of SNA-II is able to permeate the PM, the oligomers with dimensions larger than the pore membrane are unable to pass through.

According to the molecular size and size of PM pores, it appears that the ability to traverse the PM is a critical feature for lectin toxicity. Although both

SNA-I and SNA-II have demonstrated entomotoxicity against lepidopteran and hemipteran insects, research has shown that they have the capacity to induce apoptosis in the insect gut through a caspase-dependent pathway (Macedo, Oliveira and Oliveira, 2015).

7.8. SAFETY ASSESSMENT OF LECTIN

Many distinct botanical genes with insecticidal properties are being sought after by various research groups for use in biotechnological applications. The influence of the introduction of a new gene upon naturally occurring pollinators and the formation of helpful predators, on the other hand, has not been studied as extensively as the influence of the introduction of a new gene on beneficial predators.

It has been investigated if it is possible to employ lectin-assisted tri-trophic effects for pest management. There are, however, just a few findings available, and they vary depending on the lectin, the intended insect, and the bug's natural predator.

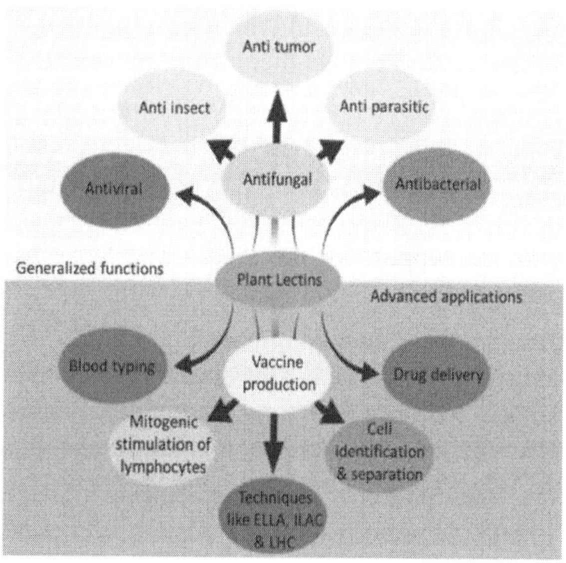

Figure 7.10. Applications of Lectins.

Source: (Mishra A, 2019)

Several studies have been conducted on the tri-trophic impact of GNA, when produced in potato plants, on the two-spot ladybirds (*Adalia*

Chapter 8

Role of Microbes in Plant-Insect Interactions

CONTENTS

This chapter tries to explain the major role of microbes that plays an important role in plant – pest interactions. This chapter also explains the microbial diversity that is involved in plant-pest interaction. This chapter also provides highlights on the various beneficial aspects of microbes-plant-pest interactions and visual cues.

Further, it also addresses the harmful aspects of microbes-plant-pest interactions and visual cues. The chapter sheds light on the role of bacteria in plant-pest interactions which includes rhizobia, and plant growth promoting bacteria. In this chapter, the role of fungal interaction has also been explained.

The chapter provides highlights on the concept of microbes and horizontal gene transfer in microbe-plant-insect interaction, microbes-insect-plant and attraction of vector, and various impact of microbes on insect reproduction.

8.1. INTRODUCTION

When it comes to food supplies, insects require sources that offer enough nutrition for their growth, development, and reproduction. It is necessary for insect herbivores to consume food plants in order to power their metabolic activities, and they must hunt for suitable host plants in a variety of ecological environments. It is common for insect herbivores to rely on plant-produced signals that offer information about key plant characteristics to find and assess the quality of potential host plants.

Plant and herbivore interactions with microbes offer an extra degree of complexity to herbivore foraging, which may be quite beneficial. The association of all plants with helpful and harmful microorganisms has been documented, and these microbes have been shown to play key roles in altering plant characteristics that impact host-plant selection by insect herbivores.

When it comes to insect herbivore foraging behavior, we describe it as the choice and placement of feeding plants, and we concentrate on research evaluating host-plant preference or colonization as the primary outcomes. This data set also includes metrics relating to herbivore oviposition preference, because the act of oviposition is a method through which gravid females select host plants for future offspring. Following herbivore interaction with microbe-associated plants, the significance of dispersion behavior and subsequent host plant selection in the development of pathogens have also been examined (De Moraes, C. et al., 2001). Most relationships between species are based on nutrition, which is the foundation of most interactions.

Insects are among the most remarkable evolutionary triumphs on the planet, with an estimated total of more than 4 million species (Novotny et al., 2002).

Herbivory is the most frequent type of feeding strategy, and the genesis of this success may be traced back to the variety of feeding methods used by these animals. Plant tissues, on the other hand, are generally nutritionally deficient due to improperly balanced ratios and/or insufficient amounts of essential nutrients, as well as the frequent requirement to detoxify plant-defensive agrochemicals.

Phytophagous arthropods must utilize a variety of pre- and post-ingestive processes to compensate for the nutritional mismatch between what plants supply and what insects require in order to be able to exploit plant resources. These strategies include specific behavioral and physiological adaptations, intricate interactions that involve insect reprogramming of host plant development, and host plant reprogramming of insect development.

For further information, check the chapter: From Plant Exploitation to Mutualism by Lieutier and colleagues (2017). There are several types of symbioses, including symbioses in which plants have evolved food rewards, particularly for insects, and relationships with one or more symbiotic partners.

Several studies have demonstrated that microorganisms are important 'hidden players' in insect-plant interactions, and that they can influence, among other traits, the insect host plant range, insect feeding efficiency, and insect metabolism. It has been demonstrated that insect symbionts can have a negative impact on the plant by interfering with plant signal transduction pathways, repressing or counteracting the expression of plant defense-related genes, or altering the primary and secondary metabolisms of plants.

Insect symbionts can also have an impact on plant-insect interactions through their direct or indirect effects on their insect host, such as by providing new metabolic pathways (and/or by altering insect reproduction or insect immunity, which can have an impact on plant exploitation. Lastly, they can regulate insect relationships with their natural enemies or plant-related organisms such as other herbivores, plant symbionts, or plant diseases (Biere& Bennett, 2013).

As insect herbivores, they have a variety of sensory systems that allow them to detect and understand information from their environment that is encoded as visual, olfactory, and gustatory signals (described in further detail below). Due to their popularity in the literature and relevance in mediating host-plant selection by insect herbivores, we will concentrate on this group of cues in this section of the chapter. The utilization of plant-associated signals by herbivores is common during foraging and oviposition, as these cues can offer information about the location, identification, nutritional quality, and defensive state of the plant. In the host-plant selection process, cues from different sensory modalities frequently play distinct roles at different stages, from the initial placement of plants or habitats to the selection of individual plants or tissues.

A vast number of insect species rely on visual cues to locate plants over long distances, which is especially true if the insects are capable of long-range dispersion. Gustatory signals, on the other hand, need direct contact with the plant and give information about appropriate tissues for eating or oviposition.

There are a variety of cues used by insect herbivorous species that differ from one another, and certain cues may be more useful in specific habitats, such as soil environments, or at specific times of the day, such as diurnally active, nocturnally active, or crepuscularly active activity.

When it comes to hosting plant selection, visual, olfactory, and gustatory signals are all important, but they are typically utilized in conjunction with hunting or ovipositing insects, depending on the species and environmental conditions.

It is common to see relationships between plants and microorganisms, which can vary from helpful to parasitic or harmful in nature. Plant-associated microorganisms are increasingly being recognized as having key functions in modifying plant phenotypes and influencing interactions between plants and insects. Growing data suggest that microorganisms modify plant-produced signals, which in turn affect the oviposition and foraging behavior of insect herbivores.

The majority of the mechanisms via which soil organisms might influence aboveground insects in the field are mediated by plants. There are a variety of organisms that can promote plant development, the most notable of which being plant growth promoting rhizobacteria (PGPR) and arbuscular mycorrhizal fungus (AMF), which has been theorized to enhance plant palatability.

Plants under biotic or abiotic stress, on the other hand, maybe more vulnerable to attack by herbivores as a result of the stress they are. Field investigations have provided evidence for the first of these hypotheses. Therefore, the nutritional value of plants (particularly their nitrogen and sugar content) in the field can be favorably impacted by soil organisms in the field. The palatability of host plants, as well as secondary defensive compounds produced by plants in response to pathogens, can be influenced by soil organisms in the field (Megas and Müller, 2010). As a result of their interactions with soil organisms, plants can become more sensitive to antagonist attacks and respond faster and with greater vigor when confronted with a future attack by an opponent.

Plant-insect interactions in the field might be complicated by a process known as induced systemic resistance (ISR). Saravanakumar et al. (2008) demonstrated that ISR can play a significant role in the field. It is also possible that soil organisms interfere with plant volatile emissions, which are crucial signals for herbivores (such as for oviposition) and many natural enemies in order to locate host plants (Megali et al., 2015). After everything is said and done, numerous studies have demonstrated that, for example, AMF can alter plant functional characteristics such as flower size and stamen number.

8.2. MICROBIAL DIVERSITY INVOLVED IN PLANT-PEST INTERACTION

Molecular methods are generally used for characterizing and identifying microorganisms associated with insects, since many symbionts cannot be grown outside of their hosts. Until the advent of next-generation sequencing, obtaining a complete or partial set of 16S or 18S rRNA genes from bacterial and eukaryotic organisms for assessment of symbiont diversity and composition was largely accomplished by cloning and sequencing.

As these genomic areas contain significant molecular differences to differentiate microbial taxa and infer symbiotic species from DNA sequences, these regions tend to be synonymous with species-level taxonomic names. In order to investigate the prevalence and transmission patterns of related microorganisms, it is necessary to create specific primers that amplify the symbionts of interest. Since these organisms belong to the same species, the multilocus sequence typing method is often used to identify strains of microorganisms that may infect various insect hosts (e.g., Henry et al., 2013).

Not only do qualitative data, but also quantitative polymerase chain reaction (PCR) primer solutions may be utilized to determine the number of symbionts within the host using quantitative PCR. When high-throughput sequencing is used to find microbial partners, instead of traditional polymerase chain reaction amplification and cloning procedures, metagenomic, genomic, and transcriptomic data are being used. The complete range of microbial communities that are linked with insects is detected, and their effect on the host phenotype is addressed using appropriate bioinformatics methods. It is now possible to get all known genome sequences for many different insect symbionts, and these whole genomes have aided in explaining the spatial processes of insect symbiosis, as well as illuminating symbiotic activities using metabolic networks inference.

Furthermore, understanding the host/symbiont relationship and variety is also necessary, as well as locating microorganisms inside their hosts in order to discover host/symbiont interactions and symbiont transmission patterns. Fluorescence in situ hybridization (FISH) is a good way to study symbionts and organs within the host which are in a symbiotic relationship.

The technological investigation, such as transmission electron microscopy, enables a more detailed examination of symbiotic ultrastructure and enables distinctions between various symbiont taxa (symbionts that reside in the same host tissue) by labeling particular antibodies (Tsuchida, Koga, Fujiwara, &Fukatsu, 2014).

Additional sophisticated technology-based research has uncovered a multitude of previously unseen microbial companions, including symbioses in plants (such as mycorrhizal and nitrogen-fixing bacterial mutualists) and insects (such as Buchneraaphidicola, the aphid obligatory symbiont).

With regards to this, insects see an increased interest in researching the ever-growing field of facultative symbionts (sometimes referred to as secondary or accessory symbionts) and, to a lesser degree, the organisms that inhabit the digestive system, which is referred to as gut associates.

Many ecological, behavioral, and physiological characteristics of their hosts may be greatly affected by the microbial communities within them, including qualities related to plant consumption, defense against natural enemies, and responsiveness to climate change (Oliver et al., 2010). In the future, whole microbial communities will be examined that exist symbiotically with plants and insects.

The plant and insect microbiomes include microbial communities that affect the host either in a surface manner or an interior manner. Next-generation sequencing (NGS) technology has enabled researchers to discover microbial diversity in various habitats and environmental situations.

These genomic datasets indicate that microbial taxa linked with their hosts are present and abundant in the environment, with certain taxa being found in low abundance. The biological functions of microorganisms may be assessed using transcriptional datasets.

However, further research is still needed to uncover the bacteria that interact with a diverse array of herbivorous insects (e.g., Lepidoptera). As the information on this project comes together, it appears that herbivorous insects have very restricted microbial ecosystems, made up mostly of a few species (Colman et al., 2012).

The typical number of taxa found in human intestines ranges from around 1000 to 3000, while Drosophila has no more than 30 operational taxonomic units (OTUs) on average. Recent papers analyzing insect species' microbial communities came to similar conclusions; microbial diversity is low (on average, 10–15 OTUs per insect) and suggested that their ecological characteristics (i.e., the process of evolution and the insects' feeding habits) are significant contributors to this microbial diversity (Colman et al., 2012).

In general, insects that eat woody plants have significantly richer gut bacterial communities, with around 103 OTUs per sample, whereas insects that feed on leaves are roughly equal in richness, with about 38 OTUs per sample. Many insects feed on plant sap, such as aphids, psyllids, and whiteflies. This group of insects has an extremely limited microbial diversity, with just three to seven operational taxonomic units per sample (Colman et al., 2012).

Most individuals in this group are connected with symbionts known as main and secondary symbionts, which are unique to distinct groups of sap feeders. While several alternative explanations for low microbial diversity have been proposed, some of these explanations may not be mutually exclusive, such as:

- An exclusive diet of phloem sap that lacks microbes;
- The host's immunity system aided by heritable symbionts; and
- Competition between symbionts and transient bacteria (Jing et al., 2014).

Plant hemipterans, which comprise species that feed on phloem, xylem, and mesophyll, together with a few species that feed on entire leaf tissues, offer an excellent system for studying the impact of diet on microbial communities.

The major microbial partners of herbivore insects are bacteria belonging to the Enterobacteriaceae family. Buchnera, the obligatory endosymbiont of aphids, as well as Carsonella, the facultative endosymbiont of psyllids,

Portiera, the facultative endosymbiont of whiteflies, and SOPE, the principal endosymbiont of rice weevils (*Sitophilus oryzae*) are some of the endosymbionts found in this large group of bacteria.

In addition to Enterobacter, acetobacter and lactobacillus are detected in Drosophila (Chandler et al., 2011). There is a very strong link between Firmicutes and Bacteroidetes and xylophagous insects, namely those that feed on rotting wood (Colman et al., 2012). One of the most frequent occupants of insects is Mollicutes, and, in particular, Spiroplasma.

There are various bacterial populations residing in different locations within the host. Insect-derived structures termed bacteriocytes house obligatory symbionts. Facultative symbionts can be found in secondary bacteriocytes, sheath cells at the periphery of main bacteriocytes, and in the hemolymph.

Many bacterial companions may be found in different niches of the insect's gut and can perform several nutritional tasks, including providing nutrients directly or influencing the insect's ability to process and assimilate food. Insects infected with bacteria like Wolbachia and Spiroplasma have their own reproductive organs altered in several ways, most notably through alterations to the host sex ratio. Mutualists that aid in bacterial reproduction and manipulators who enhance their host's fitness can be found in the same population.

Other than the above, there is an increase in research that shows bacteria may inhabit the saliva glands of insect hosts, which has major implications for plant-insect interactions, including illnesses that are spread by herbivorous insects. Also, the causative agent of Huanglongbing (citrus greening) and other citrus diseases, *CandidatusLiberibacter asiaticus*, is present in psyllid vectors (Ammar et al., 2011).

The *candidatusLiberibacter asiaticus* (other name *Salivibacter asiaticus*) has been identified in the salivary glands of mealybugs, another group of phloem sap feeders, but this strain is distinct from the one that was transmitted to plants by psyllids and does not cause illness in host plants. *Flavescence dorée*, a grapevine disease that is highly resistant to *Phytoplasma vitis*, a member of the Mollicutes, is caused by *Candidatus Phytoplasma vitis*. *C. titanus* (Cicadellidae) spreads the bacterium via Salivary Glands in its vectors and is detected in Salivary Glands in its vectors (Marzorati et al., 2006).

As an endosymbiotic bacterium (*Wolbachia*) operates through a leaf-miner caterpillar (*Phyllonorycter blancardella*), this pest modifies the phytohormonal profiles of the leaf to boost its nutritional environment (Kaiser et al., 2010).

Whereas plants are inhabited by a tremendous variety of microorganisms, such as pathogens, commensals, and possibly helpful companions, insects often only have pathogens and commensals as partners. Many plant-microbe relationships are mutually beneficial. Notable examples include mycorrhizal fungi and nitrogen-fixing bacteria, but other endophytic fungi and bacteria perform a wide range of vital activities on their host plants (Bulgarelli et al., 2013).

Usually, plant-associated bacterial communities are divided into two groups: those that reside in the rhizosphere, and those that reside in the topsoil (or the phyllosphere). In addition to plants, many other things in the environment, such as animals, viruses, and fungi, also reside in the plant phyllosphere, where microbial communities include as many as 106–107 bacterial cells cm–2 of leaf area, whether alive on or within leaves (Humphrey et al., 2014).

There is an even more diverse bacterial ecosystem residing in the root endosphere and rhizosphere (that touches the soil), which hosts an impressive 104–108 bacterial cells g–1 of root tissues and an even more impressive 106–109 bacterial cells g–1 of rhizospheric tissues (Bulgarelli et al., 2013).

More than 1,000 unique microbial species were discovered in Arabidopsis thaliana, based on pyrosequencing of the 16S rRNA gene. Yet, 21 OTUs (i.e., 67% of the total sequences) were exceptionally abundant. Few additional plant tissue types outside leaf and root tissues have been examined for microbial communities (Bulgarelli et al., 2013).

Plant vessels are the principal interface for insects that drink sap. While most plant-associated bacterial taxa include Actinobacteria, Bacteroidetes, Firmicutes, and Proteobacteria, Acidobacteria are among the most prevalent (Bulgarelli et al., 2013). Despite their similar abundance in roots and leaves,

Bacteroidetes represent the greatest share of the total population of microorganisms in the phyllosphere. When considering above- and below-ground bacterial communities, it is important to remember that their diversity and organization differ. It is assumed that separate selection mechanisms determine the difference in microbial populations between the phyllosphere and the rhizosphere.

Root exudates and root genotype are the two primary variables influencing bacterial community composition in the rhizosphere. This is the case with phyllosphere communities, which are found near the leaf surface (Bulgarelli et al., 2013). At this point, the idea requires more investigation since researchers don't know where bacterial communities originate in the phyllosphere, or how variables like insect herbivory and abiotic (e.g., moisture) conditions affect the composition of bacterial communities up and down the surface (Humphrey et al., 2014).

8.3. BENEFICIAL ASPECTS OF MICROBES-PLANT-PEST INTERACTIONS AND VISUAL CUES

Plants are commonly found to develop mutually beneficial interactions with microscopic creatures. They can be found endophytically, in plant cells, or ectophytically, based on the microbe species and the nature of their relationship with the plants. Instead, here we're interested in rhizobacteria that are involved in enhancing plant development, specifically those that foster plant growth-promoting rhizobacteria (PGPR), and fungi that are also known to foster arbuscular mycorrhizal fungi (AMF) and root endophytes.

Plant growth and pest resistance characteristics that influence insect herbivores are changed by beneficial microorganisms. When microbes like rhizobia or AMF that help plants acquire nutrients also enhance the nutritional quality of the plants for insect herbivores, the microbes become "integrators" that help increase both the nutrient acquisition and nutritional quality of the vegetation for insect herbivores. The many fungal species mentioned above, including the well-known PGPR fungus, can help boost the amount of food accessible to insect herbivores (Pineda et al., 2010).

Systemic resistance heightens plant defensive responses in contrast to helpful microorganisms, which likewise activate plant resistance via induced systemic resistance (ISR). To make plants "prepared" to develop greater defenses, ISR helps them establish faster or stronger defenses against a wide spectrum of pathogens or herbivores.

Systemic acquired resistance (SAR) is different from SAR and is due to the presence of pathogenic microorganisms. Beneficial microorganisms act in several ways, all of which result in benefits for herbivorous insects and a variety of herbivore feeding guilds or levels of specialization on certain host plants. Additionally, beneficial plant-associated microorganisms can also influence insect herbivores' feeding and oviposition behavior (Grunseich *et al.*, 2020). Insect optical sensory systems interpret visual information in the form of patterns, dimensions, and spectral quality Compound eyes, ocelli, and/or stemmata are found in insect herbivores to allow them to see visual signals, and the usage of these cues differs by species and eye shape. Visual A lot of signals rely on whether an organism is diurnal or nocturnal, which are mostly employed by above-ground creatures.

Insect herbivores utilize visual cues to both determine the location of long- and short-range plants as well as the quality of those plants. By encoding a combination of visual signals (size, shape, texture, reflectance, or colour) in the physical plant characteristics like size, shape, texture, reflectance, or colour, plant location, nutrition, and defense status may be communicated to consumers (Grunseich *et al.*, 2020).

According to this hypothesis, host-plant quality-related visual signals for insect herbivores are mainly affected by plant-associated microorganisms by either improved or reduced plant growth or biomass. Biologically relevant beneficial microorganisms have been found to boost nutrient acquisition, minimise water stress, and to promote growth-related phytohormones production (Spaepen et al., 2007) beneficial microorganisms' effect on plant development and biomass are extensively studied,

Efforts to decrease the influence of these effects on insect herbivore feeding and oviposition have remarkably few records. We exclude all of the papers that address the function of beneficial microorganisms in plant-insect interactions, in order to identify areas for prospective future study. To provide another example, the plant relationship with AMF resulted in an 87% increase in total plant biomass for seven herbaceous plant species, whereas the mass gain of African cotton bollworm was greater on plants growing in an AMF association (abstract).

AMF-induced positive growth effects may be difficult to separate from other abiotic or biotic variables. It's also unlikely that microbe-stimulated plant growth will affect plant species discrimination by all herbivores, such as grazing or in all situations. The plant biomass gained by PGPR stimulated plants was linked with lower beetle herbivore colonization in a field experiment; however the plant size most likely was not the primary driving force for these results (Grunseich *et al.*, 2020).

Research on several plant strains has previously revealed differences in insect foraging behaviour, which may be caused by differences in species or isolate of helpful bacteria. In assays in which detached leaves were used, African cotton bollworm larvae preferred plants lacking AMF, but in bioassays, they favored the biggest plants independent of AMF presence. When examining beneficial microorganisms and their effects on plant development and biomass, one must consider their effect on insect herbivores.

Herbivores are able to distinguish between plant defensive features and the size of the plant, which can impact on plant choice. It is also notable that certain insect herbivores can detect and cut plant trichomes in order to get to the leaf tissues more quickly, albeit this decreases the rate of eating and diminishes insect performance. There are many reasons why herbivores may favour plants with lower trichome concentrations to enhance foraging efficiency. Protection from such things as physical defensive structures and upkeep of microbiological defenses

Mutualisms may incur metabolic costs, which suggest there may be a possible trade-off between plants. An investigation done recently revealed that AMF-infested tomato plants had lower trichome densities and had improved herbivore effectiveness because this study was not specifically conducted on insect herbivores, it is possible that insect herbivores are able to detect a decreased investment in physical defense structures in microbe-associated plants when making foraging or oviposition decisions, which would then have a positive effect on their fitness or performance (Spaepen et al., 2007).

8.4. HARMFUL ASPECTS OF MICROBES-PLANT-PEST INTERACTIONS AND VISUAL CUES

The symptoms of plant diseases are usually caused by pathogenic microorganisms that affect plant development and/or chemistry, and they can have an impact on the performance or behaviour of insect herbivores. Plant pathogen infection, for example, frequently results in reduced plant growth, colour changes, or physical deformations of plant tissues, among other effects.

When pathogens infect plants, they can induce tissue damage that decreases photosynthate production, which, when combined with the pathogen's absorption of nutrients, might change the availability of nutrients or resources for insect herbivores (Mauck et al., 2014).

When a pathogen infects a plant, it responds by activating both physical and chemical defenses. Illness control measures can include methods to physically block or inhibit the spread of an infection, as well as the synthesis of antimicrobial chemicals to combat the pathogen (Mauck et al., 2014). A typical defensive response of plants is tailored to individual diseases, with various defense pathways or suites of defense characteristics being activated in response to phytopathogens that are either biotrophic (eating living plant tissue) or necrotrophic (feeding on dead plant tissue). Plants that have been exposed to diseases that feed on biotrophic bacteria (Biggs et al., 1987)

In most cases, systemic acquired resistance (SAR), which is a physiological condition of increased immunity against subsequent infection in distant, uninfected plant tissues, is used to boost defenses. According to whether plant defense traits against a specific pathogen also confer resistance to insect herbivores, or whether anti-herbivore defenses are suppressed as a result of cross-talk between defense pathways, plant-pathogen infection can either reduce or enhance the performance of subsequent insect herbivores (Thaler et al., 2012).

In order to establish themselves quickly and spread widely throughout plant populations, pathogenic plant microorganisms have been developed. Insects and other herbivorous arthropods, such as bees and wasps, vector certain phytopathogen species, whereas others are spread by abiotic factors such as wind or water.

Persistent, semipersistent, and nonpersistent vector-borne phytopathogens can be distinguished from each other by the type of transmission they undergo, which is determined by the length of time required for the vector to acquire and transmit the pathogen (persistent, semipersistent, or nonpersistent), as well as whether the pathogen enters the hemocoel of the vector (circulative or noncirculative) (Grunseich *et al.*, 2020).

The nature of a disease's interactions with herbivores is frequently associated with the technique used to transmit the infection. Examples include the ability of some phytopathogen species, particularly those that spread inside their vectors, to directly impact the behaviour or physiology of their vectors. Enhanced or diminished plant growth or biomass is the primary visual signals for insect herbivores that indicate the quality of the host plant.

Beneficial plant-associated microorganisms have the greatest effect on these visual cues. Plant biomass is anticipated to increase in general as a result of beneficial microorganisms improving nutrient uptake, decreasing drought stress, or producing growth-related phytohormones (Grunseich *et al.*, 2020).

It is remarkable that so little research has examined the effects of helpful microorganisms on insect herbivore feeding and oviposition, despite the fact that the effects of beneficial bacteria on plant growth and biomass have been well-documented for decades. It was shown that plant connection with AMF increased aboveground plant biomass by 87 percent in all 7 herb plant species, and the African cotton bollworm biomass gain was greater on AMF-associated plants.

According to this hypothesis, foraging or ovipositing insect herbivores could benefit from detecting plants with AMF through visual cues such as increased size in order to improve their performance or fitness, although AMF-induced positive growth effects may be difficult to distinguish from other abiotic or biotic factors in some cases. It is therefore unlikely that microbe-stimulated plant biomass increases will affect host-plant differentiation by all species of browsing herbivores or in all environments, as previously stated.

In an experimental study with cucumber plants, increased plant biomass as a result of PGPR stimulation was shown to be associated with reduced colonization of beetle herbivores, despite the fact that plant size was not likely to be the determining factor in these findings (Eigenbrode et al., 2018).

A variety of plant responses to different species or isolates of beneficial bacteria have been observed in previous research and this diversity has been shown to influence insect foraging behaviour. Different AMF isolates exhibited varied effects on several plant visual signals, including height, chlorophyll levels and leaf thickness (Roger et al., 2017), according to recent research conducted on strawberries.

In detached leaf assays, African cotton bollworm larvae favored plants that did not contain AMF; yet, in whole plant bioassays, they chose the biggest plants regardless of whether or not they contained AMF. Overall, beneficial microorganisms can have a differential effect on plant growth and biomass, which can, in turn, have a context-dependent effect on the attraction or repellence of foraging insect herbivores (Roger et al., 2017).

Herbivores may distinguish physical plant defensive structures from plant size in addition to plant size, and this can influence their choice of host plant. The ability of insects to detect and clip plant trichomes to access leaf tissues more quickly, for example, has been demonstrated. However, this behaviour has been shown to delay eating and diminish insect performance.

As a result, herbivores may choose to pick plants that produce fewer trichomes in order to enhance their foraging efficiency. Plant development of physical defensive structures, as well as the maintenance of microbial mutualisms, can result in metabolic expenditures, showing that plants may be willing to make a trade-off in order to survive.

Tomato plants colonized by AMF had reduced trichome densities and improved herbivore performance, according to a recent study. It is possible that insect herbivores may detect a reduction in investment in physical defensive positions in microbial-plants when making feeding or oviposition decisions in order to improve their performance or fitness, but this was not specifically examined in this work (Grunseich *et al.*, 2020).

8.5. BACTERIA

There are several kinds of bacteria that thrive in the soil, and they are strongly associated with plant development and quality. One well-known example of this is nitrogen-fixing rhizobia that interact with leguminous plants. Many plants growth-promoting rhizobacteria (PGPR) have been shown to produce yield, but they are also shown to prime plants for the activation of defensive mechanisms, which, in certain cases, can cause insect herbivores to become susceptible to pests (Pineda et al., 2010).

8.5.1. Rhizobia

It would be reasonable to assume that the higher quality of the plants' nutrition derived from the mutualism between plants and nitrogen fixing bacteria would benefit the insects that feed on the plants. Nevertheless, rhizobia have also been found to impact on both plant resistance responses. The results of the study, which involved the placement of potted plants in the field, showed that *Rhizobium* sp. helped boost plant protein levels in the Lima bean, *Phaseolus lunatus*, but decreased the amount of extrafloral sugar (sugar not used by the plant).

Thus, reduced visiting numbers of the related mutualist ant *Tetramorium caespitum* ensued, leading to a 75% decrease in the population. This study shows that rhizobia can interfere with the normal plant defense mechanism caused by herbivores. Cyanogenesis (a plant defensive mechanism) is enhanced in the presence of rhizobia, and this may result in less effort being spent on creating extra floral nectar for ant recruitment (Godschalx et al., 2015).

8.5.2. Plant Growth Promoting Bacteria

According to the data from the experiments, plant-mediated effects of the addition of PGPR on above ground insects in the field are consistently unfavorable. Negative interactions were found in the aboveground herbivore literature search, regardless of the insect feeding guild. *Cnaphalocrocis medialis* leaf rollers in rice fields in India were nearly halved, as four distinct *Pseudomonas fluorescens* strains were introduced to the fields as well as mixed together. It's probable that ISR is contributing to these impacts, as plants tend to upregulate their defense gene expression following exposure to herbivory.

Increase the production of plant growth-promoting rhizobacteria on above-ground herbivores. The results consistently show that the addition of PGPR to the ground has a detrimental effect on above-ground insects. All of the literature search interactions confirmed that insects feeding on leaves have no effect on aboveground herbivores rolling by the rice leaf roller *Cnaphalocrocis medialis* was reduced threefold by the addition of four distinct *Pseudomonas fluorescens* strains (Saravanakumar et al., 2008). In response to herbivory, plants that undergo bacterial treatments show greater levels of defensive gene transcription (Prabhukarthikeyan et al., 2014).

Also, inoculation with PGPR might affect the attractiveness or performance of insect predators, such as parasitic insects and predators, at higher trophic levels. Interactions showed mixed results, with half reporting negative impacts and half reporting favourable benefits.

In an experiment looking at the effects of bacterial inoculation on field-grown broccoli, the researchers discovered that the ladybug (*Coccinella septempunctata*) and various unidentified syrphid flies consistently disappeared from the inoculated plants, whereas their numbers on the non-inoculated plants remained the same (Gadhave et al., 2016). In contrast, however, plants cultivated on soils treated with *B. cereus* and *B. subtilis* had significantly greater levels of cabbage aphid parasitism by the parasitoid wasp *Diaraetiella rapae* when compared to plants grown on soils treated with *B. amyloliquefasciens* or a combination of the species (Gadhave et al., 2016).

8.6. FUNGAL INTERACTIONS

Soil fungi are a varied collection of organisms and they have been researched for many years for the role they play in above-belowground interactions. Mycorrhizal fungi are the most researched taxa due to their being present in almost all plant species. When Ectomycorrhizal fungi (EMF) develop a partnership with trees, it is called an ectomycorrhizal (EM) relationship. The relationship, which occurs across the plant world, is an ectomycorrhizal (EM) one.

Due to its poorly-understood link to the soil biota-plant-insect system, EMF has only been superficially addressed. Due to the lack of research on the impact of EMF on plant-insect interactions, however, it is unclear what impact EMF may have on insects in other directions. It was found that a five-fold increase in the population of the sap-sucking poplar aphid *Chaitophoruspopulicola* occurred on trees with EMF-treated poplar (*Populus angustifolia* x *P. fremontii*) as opposed to control trees (*Populus angustifolia x P. deltoides*) which were not exposed to EMF.

It has been demonstrated in a controlled study that AMF harms generalist chewers, but benefits specialist chewers (Koricheva et al., 2009). We learned from our literature research on chewing herbivores in plots with AMF missing that omnivores and granivores interact a lot in their natural environment (Gange and West, 1994). 83% reported neutral interactions, whereas 17% indicated a favourable plant-mediated impact on these specialists.

How chewing herbivores are affected by plant-mediated AMF impacts also varies depending on various plant functional groups. At the level of plant functional groups, the herbivory levels in AMF and control plots were similar for C3 grasses, but varied for C4 grasses and forbs(Heinen *et al.*, 2018).

In controlled experiments, sap-sucking insects show consistent benefits from AMF, and there does not appear to be any correlation between AMF's specialization and the effect of sap-sucking insects (Koricheva et al., 2009). 43% observed zero net interactions, whereas 57% reported beneficial plant-mediated impacts of AMF on sap suckers *Aulacorthum solani* populations were found to be more than tenfold higher on soybeans (*G. max*) infected with *Gigaspora margarita* than on untreated controls.

This finding coincides with the consistent results in controlled experiments, which suggest plants. There was just one research about the effect of AMF on poplar aphid populations, and it found that aphid populations were reduced by two to three times on *Populus angustifolia and Populus fremontii* trees that were planted in pots in the field (Gehring and Whitham, 2002).

It is difficult to say exactly why aphids reacted badly in this research. Despite the authors finding no substantial influence on plant performance, they also didn't find a significant change in plant chemistry. Not all plant-insect interactions may be affected by AMF activity; plant functional groupings might have different responses. Research has hitherto only been done on herbaceous species, which may result in different outcomes when applied to woody shrubs and trees. The authors examine AMF plant-insect effects on insects that are neither leaf chewers nor sap suckers : Koricheva et al. (2009). However, trends in AMF plant-insect effects on insects belonging to other feeding guilds, such as cell content feeders and leaf miners, are not easy to interpret. However, AMF addition to plants in the field had little influence on root feeders, leaf miners, and gall producers in numerous experiments.

No two generations of insects in the same research setup will have identical findings. When levels of the chemical AMF were decreased using iprodaine, it took some time for those levels to influence the fly *Chromoatomyiasyngenesiae*, the sole fly species known to mine oxeye daisy, *Leucanthemum vulgare* (Gange et al., 2003).

Also, when researchers investigated the impact of AMF species specificity on the amount of *Leucanthemum* leaves mined by *C. syngenesiae*, they found that there was a 50% increase in the biomass of the leafminer, and in plots with a higher amount of AMF, the AMF-depleted foliage was being mined at a higher rate. However, for the second generation of flies in the year of research, the substantial impacts were only discovered (Gange et al., 2005). In AMF-plant interactions, the impact on pollinating insects might be either positive or negative. According to our literature search (n = 35), 34% of the interactions found in our research were favourable, 17% were negative, and 49% indicated no impacts on pollinators. A few studies have found pollinator visits to flowers.

Since environmental factors impact pollinating insects differently at multiple levels of measurement, it is essential to take note of this variability. For example, researchers discovered that using Benomyl reduces AMF levels, and in a separate study, they investigated the impact of AMF on six commonly seen forb species (Heinen *et al.*, 2018).

While natural AMF levels do not impact the total number of pollinator visits, plots with natural AMF exhibit a 67% higher pollinator visitation rate. Visitation by large-bodied bumblebees increased by three times, while that by small-bodied pollinators like bees and flies was cut in half.

In plots with greater AMF levels, *Aster laevis* and *Solidago missouriensis* had an average of two to four times more visits by pollinators than *Aster saccharum* and *Cerastium arvensis, but Cerastium arvensis* had a total of 80% less visits by pollinators. There was no connection between AMF levels and pollinator visits to the herbs *A. millefolium*, and *Erigeron philadelphicus* (Heinen *et al.*, 2018).

8.7. MICROBES AND HORIZONTAL GENE TRANSFER IN MICROBE-PLANT-INSECT INTERACTION

Completion of the characterization is awaited. However, candidate microbial genes for herbivore-herbivore interactions are commonly found in the genomes of insect pests. These candidate microbial genes have, however, been discovered using HGT. Actually, genetic material transference is increasingly thought to have an impact on eukaryotic evolution, resulting in the acquisition of new characteristics (Boto, 2014).

Several insect-specific instances of HGT describe microbial genes that are engaged in insect plant adaptation, for example, as enzyme encoding genes that degrade and metabolize plant-derived compounds, or as detoxification genes that remove toxic plant components.

It has been shown that various herbivorous insects carry genes for Plant Cell Wall Degrading Enzymes. There are genes that are presumably endogenous insect. Gene families with horizontally acquired characteristics linked to plant eating have been particularly thoroughly researched in beetles and weevils. A particular enzyme, phylogenetically linked to the Bacillus group of enzymes, is suspected to assist with feeding within the coffee berry.

This enzyme is thought to hydrolyze galactomannan, the primary storage polysaccharide in the plant. This genetic evidence, which combines molecular and enzymatic tests, suggests that these genes were likely involved in insect adaption to plants in the past. Multiple HGT of genes that control sugar and amino acid metabolism has been seen in lepidopteran genomes and 30 candidates for HGT that include mostly carbohydrate metabolic enzymes have been found in the Hessian fly (Zhao et al., 2015).

Another mechanism by which microorganisms could have contributed to the introduction of genes into phytophagous insects is by transferring genes to them from other species. Mites and lepidoptera have been found to feed on plants and use a bacterial gene to synthesize deadly hydrogen cyanide as a byproduct.

B-fructofuranosidase genes of possible bacterial origin are said to be responsible for silkworm caterpillars' immunity to sugar mimic alkaloids in mulberry latex, which are poisonous to other insects that do not feed on mulberries (Daimon et al., 2008).

Found in several herbivorous arthropods, fungal carotenoid biosynthetic genes may have a role in the herbivorous lifestyle. HGT events in insect genomes identified in hemipteran species that host obligate endosymbiotic bacteria are possible, as it has been demonstrated in insects that possess strains of endosymbiotic bacteria that are phylogenetically distinct from those of the bacteria and those genes facilitate the cooperative relationships between the two organisms. They support nutritional symbiosis by contributing to HGT in the host plant (Grunseich *et al.*, 2020).

According to one findings , HGT-mediated gene acquisition likely has a major impact on the shift to parasitic plant-dwelling and herbivorous lifestyles, since it has been discovered in a number of other parasite species, such as mites and worms (Grunseich *et al.*, 2020).

It is predicted that as the quantity of genomic data on insects grows, this will uncover additional examples of HGT. It may also be possible to officially relate the acquisition of insect adaptive characteristics to the development of novel functional methods that allow targeted gene inactivation.

8.8. MICROBES-INSECT-PLANT AND ATTRACTION OF VECTOR

Pest-associated microorganisms (IAMs) can be harmful to plants, and it is common for these plant diseases to manipulate plant–insect interactions in order to increase their own spread. Because plant viruses may be injected into plants and then proliferate and disseminate throughout the plant's whole system, they can exert their effects on host plants for an extended length of time.

Some of these diseases not only inhibit plant defense mechanisms, but they also modify the architecture and/or physiology of the plant in order to attract vector insects, which increase the likelihood of dissemination and spread. Phytoplasmas are plant diseases that are spread by insects. They cause a variety of symptoms, including dramatic morphological alterations in the host plant.

Recent research has revealed that phytoplasmas produce effector proteins that cause these types of changes. In the aster yellow phytoplasma strain Witches' Broom (AY-WB, Candidatus (Ca.) *Phytoplasma asteris*), one of the effector proteins, SAP11, was discovered in the leafhopper *Macrosteles quadrilineatus*, which is transmitted by the aster yellow phytoplasma (Bai et al., 2009).

Sugio et al. (2011) discovered that SAP11 induces instability of a subset of Arabidopsis TCP transcription factors, which are components of the regulatory components of JA production. Because JA signaling enhances Arabidopsis protection from leafhoppers, leafhoppers generate more offspring on phytoplasma-infected and SAP11-expressing Arabidopsis plants as compared to non-infected wild-type plants in the presence of JA signaling.

SAP54 has been identified as a phytoplasma effector that induces the production of leaf-like flowers (phyllody and virescence) and a large number of stems (witches' broom) in Arabidopsis plants infected with AY-WB.

SAP54 has been identified as a phytoplasma effector that induces the production of leaf-like flowers in Arabidopsis plants infected with AY-WB. SAP54 binds to members of the MADS-domain transcription family, which plays a critical role in the control of floral development, and facilitates their degradation, according to the manufacturer (Maclean et al., 2014).

It is worth noting that Arabidopsis plants with leaf-like flowers, produced by either the SAP54 single gene or the AY-WB infection, are more resistant to pests and diseases. *M. quadrilineatus*, a leafhopper vector, finds it appealing for oviposition on the plant (Maclean et al., 2014). As a result, SAP54 permits the phytoplasma to modify the morphology of the plant, which improves the plant's attraction to the leafhopper and, therefore, increases the likelihood of the phytoplasma spreading in the wild.

It is interesting to note that an Arabidopsis ap1 mutant, who produces flowers that resemble leaves, does not appear to be any more appealing to insects than healthy plants. That the flower shape in the ap1 mutant is insufficient to attract the insect and/or that SAP54 has an impact other than the development of leaf-like flowers, such as modulation of the phytohormone-mediated defense mechanism, is suggested by this finding (Maclean et al., 2014). Two AY-WB phytoplasma effectors (SAP54 and SAP11) as well as an effector of AY-WB phytoplasma

Ca. P. asteris OY strain proteins all stimulate the development of numerous leaves and stems in Arabidopsis or *Nicotiana benthamiana* plants, creating a bushy look that is a typical sign of phytoplasma infection in these plants (Sugio et al., 2011). The untested notion is that a bushy morphology induced by the development of young sensitive tissues might make a plant more appealing to insect vectors, but this has not been tested yet.

Some plant diseases are also capable of altering the volatile content of plants in order to attract insect vectors. A phytoplasma strain, Ca. *P. mali*, alters the volatile content of apples and attracts the psyllid vector, *Cacopsyllapicta*, to the fruit (Sugio *et al.*, 2015).

Apple tree psyllids are attracted to the sesquiterpene -caryophyllene, which is produced by Ca. P. mali and enhances the emission of this sesquiterpene. In a similar vein, *Ca. Liberibacter asiaticus* alters the volatile content of infected plants in order to attract the psyllid vector *Diaphorina citri* and facilitate its spread. The release of methyl salicylate caused by the Liberibacter infection is what draws *D. citri* to the infected area (Mann et al., 2012). It has not yet been determined what methods these two diseases use to alter the volatile compounds of plants in their host plants.

Despite the fact that certain insect-associated bacteria are not transmitted from insect to plant, they may interfere with the transmission of another insect-associated microbe. Buchnera, an obligatory symbiont of aphids, has been revealed to have a chaperonin protein, GroEL, that binds to the coat proteins of the luteovirus potato leafroll virus, according to research.

Aphid haemolymph is considered to be the source of the GroEL, which is generated by Buchnera in bacteriocytes and then released. In the laboratory, antibiotic treatment of the aphid lowered the quantity of GroEL in the aphid's haemolymph and reduced viral transmission by more than 70%. Considering that aphids treated with antibiotics had their major coat protein degraded, it was hypothesised that GroEL binds to the virus and protects it from enzymatic breakdown while in the aphid haemolymph, thereby contributing to the virus's persistence in the aphid and eventual transmission to the plant. However, conflicting results have been provided in a study of the interaction between the pea aphid and another luteovirus (the barley yellow dwarf virus) (Bouvaine et al., 2011), which was conducted in the field.

The transmission of the begomovirus tomato yellow leaf curl virus by another hemipteran insect, the whitefly, has been found to be aided by a secondary symbiont in the case of this insect (TYLCV). While the Israeli whitefly B biotype, which is closely connected with Hamiltonella, is effective at transmitting the virus, the Q biotype, which is associated with other secondary bacterial symbionts, is ineffective at transmitting the virus, according to the researchers (Gottlieb et al., 2010).

It has been demonstrated that the *Hamiltonella* GroEL interacts with the TYLCV coat protein both *in vitro* and *in vivo*, and that this interaction contributes to virus transmission by the whitefly (Morin et al., 1999; Gottlieb et al., 2010), and that this interaction occurs only in the biotype that carries the *Hamiltonella* GroEL (Gottlieb et al., 2010).

It has not yet been established if insect-associated microorganisms affect the pace at which plant-pathogenic bacteria spread across the environment. A critical study field that requires greater exploration in order to better understand disease epidemiology and to enhance disease control methods is cancer immunology.

8.9. IMPACT OF MICROBES ON INSECT REPRODUCTION

The technique adopted by symbionts to maintain and invade indeed, because of this uniparental transmission, tactics that promote daughter production at the expense of male development are more likely to be adopted. In addition, such effects include the emasculation of genetic males, the implantation of thelytokous parthenogenesis, the killing of diseased offspring, and the induction of cytoplasmic incompatibility (CI), a type of postzygotic reproductive sterility that occurs when diseased males mate with non - infected females or females infected with another strain of the microscopic organism. In addition to the bacteria Wolbachia, additional symbionts that have the ability to influence their host's reproduction include Arsenophonus, Cardinium, Rickettsia, and Spiroplasma (Engelstter& Hurst, 2009).

According to Duron et al. (2008), these symbionts infect a significant number of insect species (for example, aphids). Contrary to expectations, Hemiptera have had relatively few examples of reproductive manipulation by their symbionts reported to yet. One possible explanation for this is because many species of aphids display full or cyclical parthenogenesis, which may reduce the efficacy of reproductive manipulation.

Simon and colleagues (2011), on the other hand, revealed that Spiroplasma causes male aphid death in the pea aphid. Germs that kill men can be perpetuated in their host populations if the male-killing bacteria provide an indirect benefit to infected females, such as by limiting rivalry with their brothers or by avoiding the onset of inbreeding depression.

This second aspect is particularly relevant in the case of the pea aphid, which undergoes cyclical parthenogenesis while maintaining the genotypic composition of its offspring through successive clonal generations. Indeed, if inbreeding depression is a problem, limiting inbreeding through male lethal during in the official event of sexual reproduction should provide an advantage not only at that time, and even during consequent clonal generation to generation, thereby benefiting the females who harbor the male lethals throughout the year, as well as the male lethals itself. It would be worthwhile to investigate this theory, either theoretically or by field surveys. Alternatively, Spiroplasma may give other benefits to its hosts, which would explain its persistence in pea aphid populations (Godfray et al., 2013). It has also been shown that the recent invasion of Rickettsia in the United States is related with greater female bias in the offspring of infected females which may have an impact on the population dynamics of whiteflies.

While not directly influencing insect-plant interactions, reproductive manipulators have the potential to influence population dynamics (either positively or negatively) and have been shown to reduce genetic diversity and/or recombination rates in infected species. This has the potential to have a negative influence on the functioning of ecological networks and their co-evolutionary processes as a result of cascading consequences (Ferrari &Vavre, 2011).

These impacts might have an impact on not just plant-insect relationships, prey-predator interactions, or host-parasitoid interactions, but also on competitive interactions at a particular trophic level, among other things (e.g., between herbivores). As an illustration, the virus LbFv, which manipulates the reproductive behaviour of the parasitoid *Leptopilina boulardi*, reduces the competitive ability of this species against Leptopilina. Importantly, the impacts of reproductive manipulators on ecological networks, and particularly on insect-plant interactions, have largely gone unexplored despite the fact that they have the potential to be significant.

Reproductive manipulators may potentially be a promising possibility for the development of alternative control methods in the context of reproduction. Wolbachia, in particular, may be exploited to create insect incompatible methods by inducing CI in the bacteria (IITs).

It was discovered that the fruit fly *Ceratitis capitata* (Diptera: Tephritidae) was capable of demonstrating the concept (Zabalou et al., 2004). This species has never been infected by any type of reproductive manipulator in the wild. A Wolbachia strain from Rhagoletiscerasi (Linnaeus, 1758) has been introduced into *C. capitata* by artificial transfection, and it has been shown to cause 100 per cent CI in this species.

Using population cage studies, it has been demonstrated that releasing infected males may actually be utilized to regulate host populations in a manner that is very similar to sterile insect approaches (SIT). One benefit of IIT over SIT is that infected males are typically considerably more competitive than irradiated males, which is a significant advantage in the breeding process. One disadvantage is that only males should be released, which necessitates the use of ideal sex techniques, despite the fact that alternative tactics that mix IIT and SIT have recently been presented (Sugio *et al.*, 2015).

8.10. CONCLUSION

In the conclusion of the chapter, the major role of microbes that plays an important role in plant – pest interactions have been discussed. This chapter also discussed about the microbial diversity that is involved in plant-pest interaction. In this chapter, the various beneficial aspects of microbes-plant-pest interactions and visual cues have also been discussed.

This chapter also discussed about the various harmful aspects of microbes-plant-pest interactions and visual cues. In this chapter, the role of bacteria in plant-pest interactions, which includes rhizobia, and plant growth promoting bacteria, the role of fungal interaction in plant-pest interactions has also been discussed.

Towards the end of the chapter,the concept of microbes and horizontal gene transfer in microbe-plant-insect interaction, microbes-insect-plant and attraction of vector, and various impact of microbes on insect reproduction have also been discussed.

24. Daimon, T., Taguchi, T., Meng, Y., Katsuma, S., Mita, K., & Shimada, T. (2008). Beta-fructofuranosidase genes of the silkworm, Bombyx mori: Insights into enzymatic adaptation of B. mori to toxic alkaloids in mulberry latex. The Journal of Biological Chemistry, 283, 15271e15279.

INDEX

U

V

W

X